MONOGRAPHS ON
STATISTICS AND APPLIED PROBABILITY

General Editors

D.R. Cox, D.V. Hinkley, D. Rubin and B.W. Silverman

(Full details concerning this series are available from the publishers)

Analysis of Repeated Measures

M.J. CROWDER
Senior Lecturer, Department of Mathematics,
University of Surrey

and

D.J. HAND
Professor of Statistics, The Open University

CHAPMAN & HALL
London · New York · Tokyo · Melbourne · Madras

Published by Chapman & Hall, 2–6 Boundary Row, London SE1 8HN

Chapman & Hall, 2–6 Boundary Row, London SE1 8HN, UK

Chapman & Hall, 29 West 35th Street, New York NY10001, USA

Chapman & Hall Japan, Thomson Publishing Japan, Hirakawacho Nemoto Building, 7F, 1-7-11 Hirakawa-cho, Chiyoda-ku, Tokyo 102, Japan

Chapman & Hall Australia, Thomas Nelson Australia, 102 Dodds Street, South Melbourne, Victoria 3205, Australia

Chapman & Hall India, R. Seshadri, 32 Second Main Road, CIT East, Madras 600 035, India

First edition 1990
Reprinted 1991

© 1990 M. J. Crowder and D. J. Hand

Typeset in 10/12 Times by Thomson Press (India) Ltd, New Delhi
Printed in Great Britain by St Edmundsbury Press Ltd,
Bury St Edmunds, Suffolk

ISBN 0 412 31830 X

A catalogue record for this book is available from the British Library

Library of Congress Cataloging-in-Publication Data
Crowder, M. J. (Martin J.), 1943–
    Analysis of repeated measures/M. J. Crowder and D. J. Hand.
    —1st ed.
      p. cm.—(Monographs on statistics and applied probability; 41)
    Includes bibliographical references.
    ISBN 0-412-318390-X
    1. Multivariate analysis.   I. Hand, D. J.   II. Title
III. Series.
QA278.C76   1990   90-1339
519.5′35—dc20   CIP

Contents

Acknowledgements

We thank: Alan Kimber and Linda Wolstenholme for their bravery; a referee for his valuable comments on the first draft; and the many researchers who gave us permission to use their data.

Preface

After several years of having to answer frequent enquiries about repeated measures analysis from various researchers, it occurred to me one day that I might take the trouble to find out something about the subject. In the belief that a good way of learning a topic is to have to teach it, a one-day course was arranged to be given at Surrey University in April 1987. Course notes were prepared and the presentation was gratefully shared between my intrepid colleagues, Alan Kimber and Linda Wolstenholme, and my trepid self. The day went well (it didn't rain) and the idea of a book was born. It is fairly certain that nothing more would have happened had I not met David Hand at the PSI conference in Harrogate later that year. He had also decided to write a book on repeated measures and, being the dynamic chap that he is, would soon have completed the project! I was very glad to grab the chance of collaboration with someone who already had so much experience in writing textbooks – an old Hand, one might say.

Martin Crowder
Surrey University
September 1989

Like my co-author, I had spent many years answering questions about how to analyse repeated measures experiments. They must surely be amongst the most common kinds of designs arising in statistical consultancy work. However, their frequency of occurrence is matched by the large number of different techniques which have been formulated for analysing such problems. This posed a problem – there seemed to be no concise overview of the techniques which went into sufficient detail to be useful – either for statisticians or for students or researchers who wished to apply such methods. Thus I felt that a book doing just that was necessary. I was delighted when Martin Crowder, who had had the same thoughts himself, agreed to collaborate with me on this book.

This volume is the result. It is fairly mathematical – for consultant statisticians rather than for their clients – though we have included many examples arising from our work and other sources. It will not be perfect – both of us were aware, as we were writing it, that the field is developing rapidly. It is best described as an attempt to bring together the methods which exist for analysing repeated measures problems – in response to my co-author's concluding sentence above, it is an attempt to collate a Crowder methods currently available.

David Hand

Open University

September 1989

CHAPTER 1

Introduction

1.1 Background and objectives

Repeated measurements arise in many diverse fields, and are possibly even more common than single measurements. The term 'repeated' will be used here to describe measurements which are made of the same characteristic on the same observational unit but on more than one occasion. In longitudinal studies individuals may be monitored over a period of time to record the developing pattern of their observed values. Over such a period the conditions may be deliberately changed, as in crossover trials, to study the effects on the individual. Even in studies which are not intentionally longitudinal, once a sample of individual units has been assembled, or identified, it is often easier and more efficient in practice to monitor and observe them repeatedly rather than to discard each one after a single observation and start afresh with another sample.

Repeated measures may be spatial rather than temporal. For instance, for individual load-bearing cables the breaking strength may be measured at several points along the length, so 'time' becomes 'distance'. Again, in two dimensions, the intensity of corrosion may be recorded at points over the floor area of a metal tank.

A long list could be made of fields of application of repeated measures analysis. Rather than doing this let us note that in most contexts where a single measurement can be made, repeated measurements can be made. The exceptional cases are often of the type where measurement or observation is destructive in some way. If observation only alters the state of a unit this can often be included in the repeated measurements model. For instance, in education successive test scores can incorporate a learning effect.

The basic data format may be identified tersely as '*n individuals* × *p measurements*'. The individuals may be humans, litters of animals, pieces of equipment, geographical locations, or any other unit for which the observations are properly regarded as a collection of connected measurements. The measurements on an individual are recorded values of a variable made at different times. If more than one variable is recorded at each time the data form a three-way array: *individuals × measurements × variables*.

The aim of this account is to steer the reader through a pool of techniques available for analysing repeated measurements. Starting with very simple statistical methods, which are often not very efficient in this context, it moves through intermediate methods, as commonly employed by the professional statistician, and finally gets into deeper water. Most readers will be able to wade through the shallower material and it is hoped that no one will drown in the deeper stuff later on. The latter comprises techniques based on models which try to be more realistic and comprehensive, and thus require more complicated analysis. To achieve this the emphasis will be on discussion and appraisal of the more advanced techniques, with numerical examples, rather than getting too immersed in the mathematical details.

1.2 Overview

An appropriately random selection of examples will be described in connection with the various statistical methods to be covered. The data sets are all genuine, arising from various contacts with researchers wanting their results to be processed statistically (for one reason or another). Represented here are undergraduate projects, postgraduate theses and postdoctoral research at Surrey University and London University Institute of Psychiatry, as well as pharmaceutical companies' trials and hospital consultants' research projects. However, the figures and descriptions have been disguised where necessary for reasons of confidentiality. Also, some of the data sets are simplified subsets from more extensive and complex studies. The accounts have been restricted to a bare description of the situation from which the data arose, together with the figures and some plots. These examples will be used to illustrate both the qualitative deliberations and the quantitative methodology, i.e. the chat and the sums.

Some simple analyses are described and commented upon in Chapter 2. These are techniques which are, in many cases, rather inefficient but which tend to be employed quite regularly because they involve only very simple and familiar operations. The methods here include t-tests, comparison of summary features of curves, and regression.

Anova methods are the subject of Chapter 3. This sort of approach has been developed mainly since the 1960s and is now widely implemented in computer packages. However, its routine use is limited to fairly balanced designs where there is a tidy partition of the total sum of squares, and fairly simple models which produce a suitable covariance structure for the observations.

In Chapter 4 so-called multivariate methods are described in which the covariance structure is unrestricted. In this approach, the analysis is more in line with regression than Anova. Thus, it is easy to work with both factor effects and covariates, and to examine explicitly particular parameters.

The advantages of general linear regression modelling and appropriate covariance structure are both incorporated into the approach described in Chapter 5. Also, some commonly occurring covariance structures are described in detail.

Two-stage linear models involving random regression coefficients are outlined in Chapter 6. This very general set-up received a boost by Lindley and Smith (1972), who treated it in a Bayesian context, and many ramifications and applications have appeared since in the literature.

A common special case of a repeated measures design is the crossover study. Such designs have been the focus of much recent controversy, as discussed in Chapter 7.

As elsewhere in statistics, categorical data pose their own problems for repeated measures analysis. Approaches to these are described in Chapter 8.

Finally, Chapter 9 discusses some further important topics, and Chapter 10 gives some guidance to currently available software together with some worked examples.

The level of mathematical difficulty is not uniform throughout the book, nor even monotone increasing. Some sections may be scanned lazily in the cosy fireside armchair, while others will call for sleeves to be rolled up. This just reflects the widely varying sophistication of different approaches to repeated measures. Most readers will probably find some familiar material and some less familiar material here. In any case, we hope that the gathering together of the various techniques, and the consequent cross-relating of them, will be found useful.

There is very little 'original' material here. Our inclination for barefaced plagiarization from as many sources as possible has been tempered only by the need to write a coherent account. We have tried to give credit wherever due, though, for tidiness, most references have been collected at the ends of chapters rather than scattered throughout the text.

CHAPTER 2

Some simple analyses

2.1 Comparisons at individual times

Example 2.1

Four groups of subjects took part in a medical experiment: a normal control group ($n = 8$), a group of diabetic patients without complications ($n = 6$), a diabetic group with hypertension ($n = 7$), and a diabetic group with postural hypotension ($n = 6$). The subjects performed a small physical task at time 0 and two variables were recorded at times $-30, -1, 1, 2, 3, 4, 5, 6, 8, 10, 12, 15$ (minutes). The raw data for variable 1 are given in Table 2.1, which contains several missing values coded as -9.0. Plots for the separate groups appear in Fig. 2.1 where the points at -30 minutes are positioned at -5 for convenience. It can be seen that at least two subjects (one in Group 2 and one in Group 3) eventually outstrip their fellows by a handsome margin.

The question of interest concerns differences between the groups. Every 'first course in statistics' covers the two-sample t-test, and not too much further on in most cases, so this is the technique most commonly thrown at data like these by the average investigator.

Let us focus on the time 5 minutes and Groups 1 and 2. The sample sizes are $n_1 = 8$ and $n_2 = 6$, the sample means are $\bar{y}_1 = 6.81$ and $\bar{y}_2 = 6.98$, and the sample variances are $s_1^2 = 12.30$ and $s_2^2 = 12.79$. Hence the pooled sample variance is $s^2 = 12.50$ and the t-statistic is $t_{12} = (\bar{y}_1 - \bar{y}_2)/s(n_1^{-1} + n_2^{-1})^{1/2} = -0.9$. This small value will not persuade anyone of gross differences between the groups. Here, as elsewhere, we will assume that the necessary assumptions for the standard method are met. In the present case these are that (i) the observations form independent samples from normal distributions, and (ii) the population variances are equal.

Continuing in this vein we may test between Groups 1 and 3,

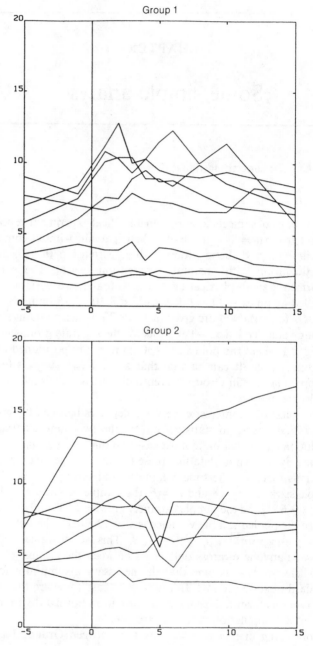

Fig. 2.1. *Effect of a physical task on hospital patients.*

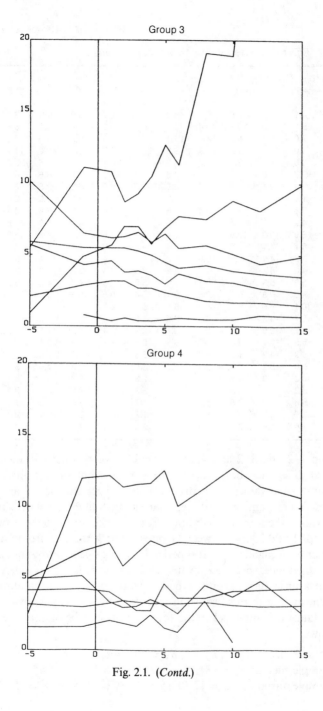

Fig. 2.1. (*Contd.*)

Table 2.1. *Effect of a physical task on hospital patients*

Subject		Time (minutes) -30	-1	1	2	3	4	5	6	8	10	12
Group 1	1	4.1	6.1	7.6	7.5	8.9	9.5	8.7	8.8	-9.0	7.0	-9.0
	2	5.8	7.5	10.1	10.4	10.4	8.9	8.9	8.4	9.9	8.6	-9.0
	3	7.0	8.4	11.2	12.8	10.0	10.3	9.5	9.2	9.0	9.4	-9.0
	4	9.0	7.8	10.8	10.3	9.3	10.3	11.5	12.3	10.0	11.4	-9.0
	5	3.6	4.3	3.9	3.9	4.5	3.2	4.1	4.0	3.5	3.7	3.0
	6	7.7	7.0	6.7	7.0	7.9	7.4	7.3	7.2	6.6	6.6	8.3
	7	3.4	2.1	2.2	2.0	2.2	2.2	2.5	2.3	2.5	2.4	2.0
	8	1.8	1.4	2.1	2.4	2.5	2.3	2.0	2.0	1.9	2.0	2.0
Group 2	9	7.6	8.9	8.5	8.4	8.5	8.2	5.6	8.8	8.8	8.4	8.0
	10	4.2	6.5	7.5	7.1	7.2	7.0	5.0	4.2	6.9	9.5	-9.0
	11	6.9	13.3	12.9	13.5	13.4	13.1	13.6	13.1	14.8	15.3	16.1
	12	8.1	7.4	8.8	9.2	8.4	9.2	7.9	7.9	7.9	7.3	-9.0
	13	4.5	4.9	5.5	5.6	5.2	5.3	6.4	6.0	6.4	6.4	-9.0
	14	4.2	3.2	3.2	4.0	3.2	3.4	3.4	3.2	3.2	3.2	2.8
Group 3	15	5.9	5.5	5.5	5.5	5.3	5.0	4.5	4.1	4.3	3.9	3.7
	16	-9.0	0.8	0.4	0.6	0.4	0.4	0.5	0.6	0.5	0.5	0.8
	17	10.1	6.5	6.2	6.3	6.6	5.9	6.5	5.5	5.7	5.1	4.4
	18	5.7	4.3	4.6	3.8	3.9	3.6	3.0	3.7	3.2	3.1	2.7
	19	2.1	2.9	3.2	3.2	2.7	2.7	2.4	2.2	1.8	1.7	1.7
	20	5.5	11.1	10.8	8.7	9.3	10.5	12.7	11.3	19.1	18.9	37.0
	21	0.9	4.9	5.7	7.0	7.0	5.8	6.9	7.7	7.5	8.8	8.1
Group 4	22	5.0	5.2	3.4	3.0	3.1	3.6	3.2	2.6	4.6	3.8	4.9
	23	4.2	4.3	4.1	3.5	2.8	2.8	4.7	3.7	3.7	4.2	-9.0
	24	3.2	3.0	3.3	3.5	3.4	3.3	3.3	3.3	3.4	3.2	3.1
	25	5.0	6.9	7.5	5.9	-9.0	7.7	7.3	7.6	7.5	7.5	7.0
	26	2.5	12.0	12.2	11.4	11.6	11.7	12.6	10.1	11.4	12.8	11.5
	27	1.6	1.6	2.1	1.9	1.7	2.5	1.6	1.3	3.5	0.6	-9.0

1 and 4, 2 and 3, 2 and 4, and 3 and 4. The t-values will measure scaled differences between the group means but the simultaneous quoting of p-values is incorrect. This is because the t's are not independent, and so the separate probabilities do not simply determine the joint probabilities. For instance, in the tests between Groups 1 and 2, and between Groups 1 and 3, the data from Group 1 enter both calculations. It is perfectly acceptable to quote standardized differences between means, such as $(\bar{y}_1 - \bar{y}_2)/s$, for reference, but it is misleading to convert these to t's and quote corresponding p's.

One answer to this problem is provided by one-way Anova. Thus the data at 5 minutes for the four groups can be summarized as follows.

sample sizes: $n_1 = 8$, $n_2 = 6$, $n_3 = 7$, $n_4 = 6$;
sample means: $\bar{y}_1 = 6{\cdot}81$, $\bar{y}_2 = 6{\cdot}98$, $\bar{y}_3 = 5{\cdot}21$, $\bar{y}_4 = 5{\cdot}45$;
sample variances: $s_1^2 = 12{\cdot}30$, $s_2^2 = 12{\cdot}79$, $s_3^2 = 15{\cdot}96$, $s_4^2 = 15{\cdot}94$.

In time-honoured fashion one can now compute the sum of squares between groups as $ssB = \sum n_i(\bar{y}_i - \bar{y})^2 = 16\cdot73$, with three degrees of freedom, and the sum of squares within groups as $ssW = \sum(n_i - 1)s_i^2 = 325\cdot51$, with 23 d.f.; here $\bar{y}$ denotes the overall mean $n^{-1}\sum n_i\bar{y}_i = 6\cdot13$ where $n = \sum n_i = 27$. Then the variance ratio test statistic is $F_{3,23} = 0\cdot39$, a small value not indicating strong differences between groups. It is assumed for the validity of the test that the data within groups are normally distributed, each group having the same variance.

There is a refinement of the above analysis which is commonly employed. This is to subtract the basal levels first for each individual. For instance, taking the time -30 minutes as basal, the value $8\cdot7$ for subject 1 at 5 minutes is adjusted to $8\cdot7 - 4\cdot1 = 4\cdot6$. Then the computations proceed with the adjusted values. Whether -30 minutes or -1 minutes, or some average of the two, is taken as basal is a medical decision. The adjustment is a reflection of the widespread biological observation that different individuals' values tend to be at different overall levels, and that it is therefore necessary to make each subject 'serve as his own control'. This is quite evident here in Fig. 2.1. It will be shown in later chapters how such variation can be naturally incorporated into more comprehensive statistical modelling. □

Example 2.2

Two drug treatments, both in tablet form, were compared using five volunteer subjects in a pilot trial. There were two phases, with a wash-out period in between. In each phase the subject yielded a blood sample at times 1, 2, 3 and 6 hours after medication, and the antibiotic serum levels are given in Table 2.2. Figure 2.2 displays the time-profiles for individual subjects after receiving each drug. Differences between the drugs are not visually well defined here if one is

Table 2.2. *Pilot trial for two drugs*

	Drug A				Drug B			
Subject	Time (*hours*)				Time (*hours*)			
	1	2	3	6	1	2	3	6
1	1.08	1.99	1.46	1.21	1.48	2.50	2.62	1.95
2	1.19	2.10	1.21	0.96	0.62	0.88	0.68	0.48
3	1.22	1.91	1.36	0.90	0.65	1.52	1.32	0.95
4	0.60	1.10	1.03	0.61	0.32	2.12	1.48	1.09
5	0.55	1.00	0.82	0.52	1.48	0.90	0.75	0.44

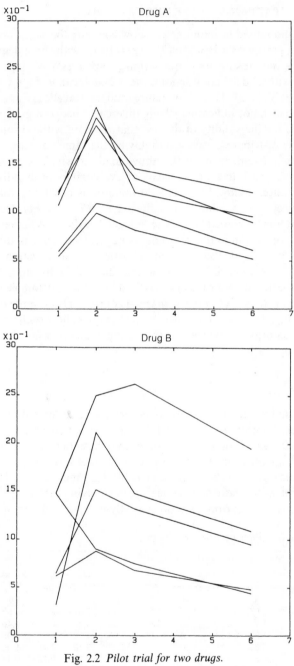

Fig. 2.2 *Pilot trial for two drugs.*

looking for consistency among the subjects. However, in common with personal beer consumption and public weather forecasts, the high is followed by a low in all cases.

The same sort of analysis can be performed as described above for Example 2.1 but now the data come in matched pairs. Let us examine the 2-hour figures, the differences Drug A − Drug B being -0.51 ($= 1.99 - 2.50$), 1.22, 0.39, -1.02 and 0.10. Thus $n = 5$, $\bar{d} = 0.036$ and $s_d^2 = 0.736$ and so the t-value for testing a zero mean difference is $t_4 = 0.09$ ($p = 0.93$). □

Comments. The above analyses have concentrated on the measurements made at particular times, at 5 minutes in Example 2.1 and at 2 hours in Example 2.2. In view of the investigator's time, effort and aspirations it might be thought just a trifle insensitive of the statistician to ignore most of his data. Thus, the analyses might simply be repeated for each time point; thus, Example 2.2 would yield four paired t-tests, one for each of the times 1, 2, 3 and 6 hours. However, such repeated tests do not in general provide a useful description of the difference between response curves. Further, the problem of dependence between test statistics now sidles in, and in a more subtle way than discussed in Example 2.1, where the same observations entered different statistics. It is now because, under reasonable assumptions about individual data profiles, there is dependence between observations at different times. For instance, consider t-testing between Groups 1 and 2 for times 5 and 10 minutes in Example 2.1. Suppose that the level for subject 1 at 5 minutes were 18.7 instead of 8.7. Obviously the t-test at 5 minutes would be affected, but also we would be alerted to the likelihood of subject 1 being a high-scoring individual. If the hypothesis about varying individual levels (referred to above) is true, then the observation at 10 minutes is now also more likely to be high and, if so, a potential t-test at 10 minutes will be affected. Thus the two t-values are linked in probability via the 5- and 10-minute figures, which are not independent. This phenomenon will appear more clearly in terms of the formal models derived below. To sum up, repeated t-tests are usually invalid.

2.2 Response feature analysis

In practice there is (or should be) particular interest in certain well-defined aspects of the data, these being identified prior to the statistical analysis. For instance, considering the trace of a body function following some temporary activation, vital information might reside in the initial phase (maximum rate of increase, time to achieve this, etc.), the middle phase (peak value, time to peak, curvature there), and the later return to equilibrium (half-life, maximum fall rate). In other situations interest might be centred on certain linear combinations of measurements, e.g. the average of the post-treatment levels minus that of the pre-treatment ones. Again, in the absence of firm prior definition, traditional linear, quadratic and higher-order contrasts can be examined (e.g. Rowell and Walters, 1976; Walters and Rowell, 1982). Yates (1982) gives a concise guide to this approach, and some related wisdom for good measure.

The identified aspect of study, called a response feature here, can be analysed in univariate fashion since one has now just a single response from each individual. When several such features are extracted the usual practice is to perform a separate analysis for each but recognizing that the results are not strictly independent. One could subject the several features simultaneously to a multivariate analysis (see Chapter 4) but then the interpretations can become less clear. There follows now an example of a commonly studied feature, namely the area under the curve.

In a very common type of pharmacological experiment a substance is introduced into the body, orally or by injection or infusion, and blood or urine samples are analysed at various subsequent times. Pharmacokinetic theory concerns the mechanisms of absorption and elimination of the substance by the body. A simple, commonly applied model considers the body to consist of a small number of interconnecting compartments, e.g. the blood system, the muscles, the gut, etc. In so-called **first-order kinetics** the instantaneous rate of exchange between compartments of a substance is in direct proportion to the difference in concentrations at the interface. As in the basic prototype differential equation, $dc/dt \propto c$ ($c =$ concentration, $t =$ time), this leads to solutions in which c is an exponential function, or the sum of several exponential functions, in time. The solution curve typically rises fairly steeply from the basal level after time 0, and then returns in a gentler asymptotic fashion. For such solutions, and for compartments in series, it can be shown that the total **area under the curve** (AUC) is inversely proportional to the elimination rate constant of the substance. Thus, with interest keenly centred upon the elimination rate, the AUC is analysed. For details the

particularly keen may look in Gibaldi and Perrier (1975). At a more
empirical level, the AUC is often analysed as an interpretable feature
without any appeal to underlying pharmacokinetic theory.

Example 2.3

Blood glucose levels were recorded for six volunteer subjects after
the consumption of a certain test meal. Hordes of Surrey University
students (and quite a few of us staff) had to be beaten back with
clubs at the door after word had got round about the free meals. Ten
recordings were made, at times $-15, 0, 30, 60, 90, 120, 180, 240, 300$
and 360 minutes relative to feeding time. The whole process was
repeated six times with the meals taken at various times of the day
and night. The question primarily concerns the time-of-day effect.
Table 2.3 presents the results with just a couple of missing values
(coded as $-1\cdot0$). The corresponding graphs in Fig. 2.3 show some
differences, though with much variation as usual.

To perform an AUC analysis we must first obtain the individual
AUCs. But here we run up against the first problem, i.e. what to use
for the basal value. At the present simple level let us just inspect the
data and use a rough-and-ready approximation as follows. The basal
level will be taken as the average of the last three observations, at
times 240, 300 and 360 minutes. This will then be subtracted from
the preceding values with upward adjustment to zero if necessary;
without such, admittedly arbitrary, adjustment it would be possible
in theory to end up with negative areas. Thus, for subject 1 at 10
a.m. the basal value is estimated as $(3\cdot99 + 4\cdot15 + 4\cdot41)/3 = 4\cdot18$
and the set of differences (y-basal) for times $0, 30, \ldots, 180$ is
$(0\cdot32, 3\cdot66, 1\cdot28, 0\cdot90, 0\cdot14, -0\cdot27)$; the $-0\cdot27$ is replaced by $0\cdot00$. The
second problem is how to use these adjusted differences to estimate
the AUC. Again, at this simple level we will apply the trapezoid rule.
Thus the area between 30 and 60 minutes is approximated as
$(60 - 30) \times (3\cdot66 + 1\cdot28)/2 = 74\cdot1$. Similarly, the set of areas for time
intervals 0–$30, \ldots, 120$–180 is $(59\cdot7, 74\cdot1, 32\cdot7, 31\cdot2, 4\cdot2)$, the last being
calculated as $(180 - 120) \times (0\cdot14 + 0\cdot00)/2$. The total AUC for subject
1 at 10 a.m. is then $201\cdot9$, and the full set of AUCs for the
six subjects at 10 a.m. is $(201\cdot9, 199\cdot2, 230\cdot4, 122\cdot2, 104\cdot4, 227\cdot6)$. The
corresponding AUCs at 10 p.m. are $(932\cdot8, 452\cdot4, 440\cdot1, 991\cdot9, 585\cdot3,$
$366\cdot3)$. Finally, a paired t-test between the AUCs at 10 a.m. and
10 p.m. yields $t_5 = 3\cdot65$ ($p = 0\cdot015$). $\square$

Table 2.3. *Blood glucose levels after a test meal (just after time 0)*

| | Subject | Time (minutes) | | | | | | | | | | |
		-15	0	30	60	90	120	180	240	300	360	420
10a.m. meal	1	4.90	4.50	7.84	5.46	5.08	4.32	3.91	3.99	4.15	4.41	
	2	4.61	4.65	7.90	6.13	4.45	4.17	4.96	4.36	4.26	4.13	
	3	5.37	5.35	7.94	5.64	5.06	5.49	4.77	4.48	4.39	4.45	
	4	5.10	5.22	7.20	4.95	4.45	3.88	3.65	4.21	-1.00	4.44	
	5	5.34	4.91	5.69	8.21	2.97	4.30	4.18	4.93	5.16	5.54	
	6	5.24	5.04	8.72	4.85	5.57	6.33	4.81	4.55	4.48	5.15	
2p.m. meal	1	4.91	4.18	9.00	9.74	6.95	6.92	4.66	3.45	4.20	4.63	
	2	4.16	3.42	7.09	6.98	6.13	5.36	6.13	3.67	4.37	4.31	
	3	4.95	4.40	7.00	7.80	7.78	7.30	5.82	5.14	3.59	4.00	
	4	3.82	4.00	6.56	6.48	5.66	7.74	4.45	4.07	3.73	3.58	
	5	3.76	4.70	6.76	4.98	5.02	5.95	4.90	4.79	5.25	5.42	
	6	4.13	3.95	5.53	8.55	7.09	5.34	5.56	4.23	3.95	4.29	
6a.m. meal	1	4.22	4.92	8.09	6.74	4.30	4.28	4.59	4.49	5.29	4.95	4.34
	2	4.52	4.22	8.46	9.12	7.50	6.02	4.66	4.69	4.26	4.29	4.47
	3	4.47	4.47	7.95	7.21	6.35	5.58	4.57	3.90	3.44	4.18	4.32
	4	4.27	4.33	6.61	6.89	5.64	4.85	4.82	3.82	4.31	3.81	2.92
	5	4.81	4.85	6.08	8.28	5.73	5.68	4.66	4.62	4.85	4.69	4.86
	6	4.61	4.68	6.01	7.35	6.38	6.16	4.41	4.96	4.33	4.54	4.81
6p.m. meal	1	4.05	3.78	8.71	7.12	6.17	4.22	4.31	3.15	3.64	3.88	
	2	3.94	4.14	7.82	8.68	6.22	5.10	5.16	4.38	4.22	4.27	
	3	4.19	4.22	7.45	8.07	6.84	6.86	4.79	3.87	3.60	4.92	
	4	4.31	4.45	7.34	6.75	7.55	6.42	5.75	4.56	4.30	3.92	
	5	4.30	4.71	7.44	7.08	6.30	6.50	4.50	4.36	4.83	4.50	
	6	4.45	4.12	7.14	5.68	6.07	5.96	5.20	4.83	4.50	4.71	
2a.m. meal	1	5.03	4.99	9.10	10.03	9.20	8.31	7.92	4.86	4.63	3.52	4.17
	2	4.51	4.50	8.74	8.80	7.10	8.20	7.42	5.79	4.85	4.94	2.07
	3	4.87	5.12	6.32	9.48	9.88	6.28	5.58	5.26	4.10	4.25	4.55
	4	4.55	4.44	5.56	8.39	7.85	7.40	6.23	4.59	4.31	3.96	4.05
	5	4.79	4.82	9.29	8.99	8.15	5.71	5.24	4.95	5.06	5.24	4.74
	6	4.33	4.48	8.06	8.49	4.50	7.15	5.91	4.27	4.78	-1.00	5.72
10p.m. meal	1	4.60	4.72	9.53	10.02	10.25	9.29	5.45	4.82	4.09	3.52	3.92
	2	4.33	4.10	4.36	6.92	9.06	8.11	5.69	5.91	5.65	4.58	3.79
	3	4.42	4.07	5.48	9.05	8.04	7.19	4.87	5.40	4.35	4.51	4.66
	4	4.38	4.54	8.86	10.01	10.47	9.91	6.11	4.37	3.38	4.02	3.84
	5	5.06	5.04	8.86	9.97	8.45	6.58	4.74	4.28	4.04	4.34	4.35
	6	4.43	4.75	6.95	6.64	7.72	7.03	6.38	5.17	4.71	5.14	5.19

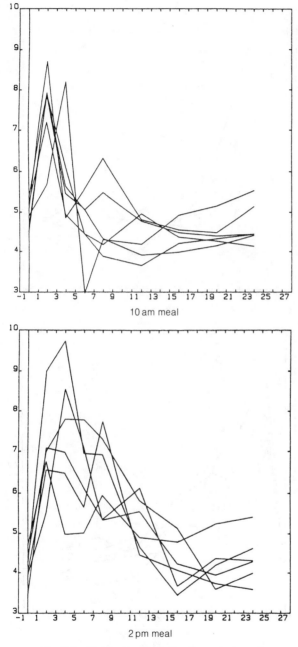

Fig. 2.3. *Blood glucose levels after a test meal.*

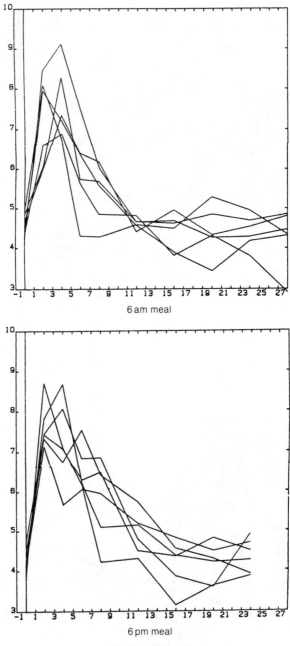

6 am meal

6 pm meal

Fig. 2.3. (*Contd.*)

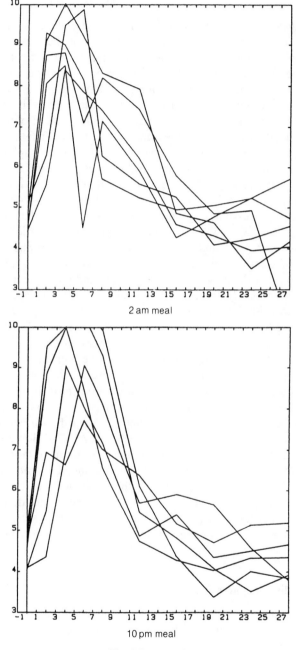

2 am meal

10 pm meal

Fig. 2.3. (*Contd.*)

Comments. The AUC analysis as implemented in the example above suffers from various deficiencies. The first is in the adoption of a basal level. This was done in a rather arbitrary manner, e.g. the value for −15 minutes was not used. The second is in the use of the trapezoid rule. In effect, the curve is replaced by straight-line segments which, for all but the very early and very late times, will lie below the curve. Thus, one is likely to obtain an underestimate of the AUC even if the data values were measured without error. This bias could possibly be reduced by the application of a more sophisticated technique such as Simpson's rule, which uses quadratic in place of linear interpolation. A third deficiency is that the curve is assumed to have returned to the basal value by time 180 minutes (240 minutes for 10 p.m.), thus ignoring the residual area beyond that point. A fourth is that when the timescale is large, small variations in the later y-values yield unduly large changes in the computed AUC. These deficiencies can sometimes be overcome by fitting a suitable curve to all of the points for an individual, and then calculating the AUC theoretically from the fitted curve. Such analyses will be described below (section 9.4).

A more fundamental doubt about AUC analysis than its implementation is whether it should be used at all, at least in isolation. The AUC is only a single feature of a curve with many potential aspects of interest. To name but three, there are the peak value, the time to attain the peak, and the half-life from peak back down to basal. Curves of very different appearance can have the same AUC. It may be argued that if one is solely concerned with elimination rates, then these are encompassed by the AUC's. However, this argument assumes that the simplest pharmacokinetic model holds true and that the observations actually obtained are true measurements of the quantity appearing in the pharmacokinetic equation. In any case, it goes against the grain to reduce a lot of data (10 or 11 points per individual in the example) to a single value, possibly throwing away much of the information, but this is probably just a personal view.

2.3 Individual curve fitting

In the last section the concept of individual curves was touched upon. In fact, in dealing with repeated measurements, it is virtually unavoidable.

Example 2.4

In a nutrition study three groups of rats were put on different diets and, after a settling-in period, their body weights (in grams) were recorded weekly over a period of nine weeks. Further, a treatment was applied during the sixth week of recording, so an effect is to be sought after that date (not before, since the plans were kept secret from the rats). This explains the extra observation taken on day 44 appearing in Table 2.4. In Fig. 2.4 differences between groups are very evident, and one can discern a setback after week 6 in Groups 2 and 3.

The change of regime during week 6 will be ignored for the time being. The individual plots in Fig. 2.4 look quite linear and fairly parallel, but the individual levels appear to differ markedly. The scientific explanation for this is that there were some big rats and some small rats. Initially then, let us fit a straight line $y = \alpha + \beta x$ to each individual profile, x denoting the day, α the intercept, and β the slope of the line. The results are given in Table 2.5: these comprise the mean $\bar{y}$, $T_{xy} = \sum(x_j - \bar{x})(y_j - \bar{y})$, $T_{yy} = \sum(y_j - \bar{y})^2$ and the least-squares estimates $\hat{\beta} = T_{xy}/T_{xx}$, $\hat{\alpha} = \bar{y} - \hat{\beta}\bar{x}$ together with the residual sum of squares RSS $= T_{yy} - \hat{\beta}^2 T_{xx}$. Since there are no missing y-values the number of time points is $p = 11$, the same for each

Table 2.4. *Rat body weights and diet supplement*

Rat		Day 1	8	15	22	29	36	43	44	50	57	64
Group 1	1	240	250	255	260	262	258	266	266	265	272	278
	2	225	230	230	232	240	240	243	244	238	247	245
	3	245	250	250	255	262	265	267	267	264	268	269
	4	260	255	255	265	265	268	270	272	274	273	275
	5	255	260	255	270	270	273	274	273	276	278	280
	6	260	265	270	275	275	277	278	278	284	279	281
	7	275	275	260	270	273	274	276	271	282	281	284
	8	245	255	260	268	270	265	265	267	273	274	278
Group 2	9	410	415	425	428	438	443	442	446	456	468	478
	10	405	420	430	440	448	460	458	464	475	484	496
	11	445	445	450	452	455	455	451	450	462	466	472
	12	555	560	565	580	590	597	595	595	612	618	628
Group 3	13	470	465	475	485	487	493	493	504	507	518	525
	14	535	525	530	533	535	540	525	530	543	544	559
	15	520	525	530	540	543	546	538	544	553	555	548
	16	510	510	520	515	530	538	535	542	550	553	569

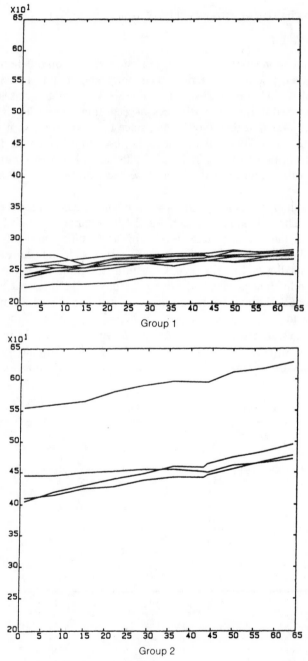

Fig. 2.4. *Rat body weights and diet supplement.*

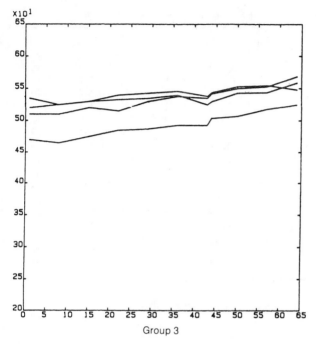

Group 3

Fig. 2.4. (*Contd.*)

individual, and $\bar{x} = 33.545$ and $T_{xx} = \sum(x_j - \bar{x})^2 = 4162.73$ for each. The fitted lines, judging by the $\hat{\alpha}$ and $\hat{\beta}$ values, vary quite a lot within groups. This can be tested more formally as follows. If the 88 points in Group 1 are thrown together and a single line is fitted, one obtains the results given in the first line of Table 2.6. If the common line fits the data almost as well as the individual lines then the overall RSS of 11 990 should not greatly exceed the sum of the individual RSSs, which is $108{\cdot}1 + 76{\cdot}2 + \cdots + 179{\cdot}4 = 960{\cdot}7$. Obviously it does, by no less than $11\,990 - 960{\cdot}2 = 11\,029{\cdot}3$, the formal variance ratio test being $F_{14,72} = (11\,029{\cdot}3/14) \div (960{\cdot}7/72) = 59{\cdot}04\,(p < 0{\cdot}001)$. This test is based on the 'extra sum of squares' rule as described by Draper and Smith (1966, section 2.7). For Groups 2 and 3 the corresponding variance ratios are $F_{6,36} = 1542{\cdot}20$ and $F_{6,36} = 83{\cdot}99$. These tests bear out one's visual impressions from Fig. 2.4 that the individual lines do not coincide. Further, from the plots, it appears that the major differences are among the intercepts, the slopes being possibly equal within groups. To test this in Group 1 eight parallel lines can be fitted by least-squares as follows. The common slope estimate

Table 2.5. *Individual regressions for rat growth data*

	Rat	$\bar{y}$	T_{xy}	T_{yy}	$\hat{\alpha}$	$\hat{\beta}$	RSS
Group 1	1	261·1	2016	1085	245	0·48	108·1
	2	237·6	1375	531	227	0·33	76·2
	3	260·2	1657	738	247	0·40	78·1
	4	266·5	1372	527	255	0·33	74·7
	5	269·4	1689	781	256	0·41	95.2
	6	274·7	1326	504	264	0·32	82·0
	7	274·6	840	437	268	0·20	267·0
	8	265·4	1701	875	252	0·41	179·4
Group 2	9	440·8	4207	4424	407	1·01	171·7
	10	452·7	5583	7604	408	1·34	117·3
	11	454·8	1509	734	443	0·36	186·6
	12	590·4	4777	5659	552	1·15	176·2
Group 3	13	492·9	3824	3703	462	0·92	190·9
	14	536·3	1310	982	526	0·31	569·7
	15	540·2	2053	1248	524	0·49	235·2
	16	533·8	3769	3648	503	0·91	235·0

Table 2.6. *Group regressions for rat growth data*

	n	T_{xx}	T_{xy}	T_{yy}	$\hat{\alpha}$	$\hat{\beta}$	RSS
Group 1	88	33302	11977	16300	252	0·36	11990
Group 2	44	16651	16076	183693	452	0·97	168200
Group 3	44	16651	10956	25669	504	0·66	18460

is given by $\hat{\beta}_c = \sum T_{xy}^{(i)} / \sum T_{xx}^{(i)}$, the individual intercept estimates are then $\hat{\alpha}_i = \bar{y}_i - \hat{\beta}_c \bar{x}_i$, and the overall residual sum of squares is $\sum [T_{yy}^{(i)} - \hat{\beta}_c^2 T_{xx}^{(i)}]$; i here refers to the ith individual, and these formulae are given in Draper and Smith (1966, Chapter 1, Exercise D). Taking the figures from Table 2.4 we find that $\hat{\beta}_c = 0.36$, the set of $\hat{\alpha}_i$'s is (249, 226, 248, 254, 257, 263, 263, 253) and the overall RSS is 1171·7. The formal test for parallelism is now based on the difference between 1171·7 and 960·7, i.e. 211·0: thus, $F_{7,72} = (211·0/7) \div (960·7/72) = 2·26$ ($0·025 < p < 0·05$). Hence, the hypothesis of parallelism looks a bit dubious. □

Comments. Whether or not one accepts parallelism, i.e. homogeneity of β's within groups, the α's certainly do vary considerably. Thus we have a problem. We want to assess differences between groups, but since there are already differences between different individuals in the same group there must automatically be differences between different individuals in different groups. So what is it that we want to assess? In the Anova tradition one thinks naturally of partitioning the variation in (α, β) into a between-groups component and a residual within-groups component. This is basically what Wishart (1938) suggested. To illustrate, the group-means of the $\hat{\beta}$'s in Table 2.1 are $\bar{\beta}_1 = 0.3600$, $\bar{\beta}_2 = 0.9650$, $\bar{\beta}_3 = 0.6575$ and the overall mean is $\bar{\beta} = 0.5856$. Thus the between-groups sum of squares is $\text{ssB} = \sum n_k(\bar{\beta}_k - \bar{\beta})^2 = 1.004$. The sums of squares of the $\hat{\beta}$'s in the three groups are 0.050, 4.231 and 0.281 so the within-groups sum of squares is $\text{ssW} = 4.562$. Hence, the variance ratio test yields $F_{2,13} = (\text{ssB}/2) \div (\text{ssW}/13) = 1.43$. This value is not large, though we would expect a more striking result from the same analysis performed on the α's.

However, this F-test requires closer examination. Let us adopt the usual framework in which the y's for the ith rat are normally distributed about the line $\alpha_i + \beta_i x$ with variance σ^2. Standard theory then yields a distribution $N(\beta_i, \sigma^2/T_{xx})$ for $\hat{\beta}_i$. If the β_i's are equal within groups then the F is valid for testing possible group differences since, fortunately in this case, the variances σ^2/T_{xx} are equal. (The variances will differ in general, e.g. when the x-values differ between individuals.) But what if we wish to allow for natural variation in β between individuals even in the same group? (Such allowance would be more or less compulsory when we came to analyse the α's.) In that event the $\hat{\beta}$'s are not identically distributed within groups, i.e. they have different means. Hence the F-test above, though based on a sensible argument, would seem not to be strictly valid. There are other puzzles here too. Should the individual estimates of σ^2/T_{xx} be pooled and used to construct the denominator of F, rather than the ssW as above? If one regards $\hat{\beta}_i$ simply as a compound measurement on the ith individual, like the AUC, how can a perfectly standard F-test be invalid?

The resolution of these difficulties will appear when the problem of naturally varying parameters is tackled in a more fundamental way. This forms the substance of the rest of this account.

2.4   Further reading

The material in section 2.1, comprising t-tests and one-way Anova, can be found in most textbooks on basic statistical methods. The analysis described in section 2.2, while elementary and widely applied in practice, seems more bashful. It is less often exposed in print, but tends to be passed on more discreetly by word of mouth.

The fitting of individual curves, and subsequent analysis of the estimated coefficients, seems to have been started by Wishart (1938). Yates (1982) is required reading. Kenward (1985, section 2) describes an application of this technique en route to bigger things.

CHAPTER 3

Univariate analysis of variance

3.1 The fundamental model

Denote by y_{ij} the jth measurement made on the ith individual. The general basis for many analyses is the partition

$$y_{ij} = \mu_{ij} + \alpha_{ij} + e_{ij} \tag{3.1.1}$$

of y_{ij} into three components:

1. μ_{ij} represents the mean level of y_{ij} over hypothetical repetitions of the set-up with randomly selected individuals from the population. Thus μ_{ij} would be unchanged if any other individual were substituted for the ith one in our sample.
2. α_{ij} represents the consistent departure of y_{ij} from μ_{ij} for the particular individual actually appearing as the ith in our sample. Thus, under hypothetical repetitions with the same individual, y_{ij} has mean $\mu_{ij} + \alpha_{ij}$.
3. e_{ij} is the 'error' term representing the departure of y_{ij} from $\mu_{ij} + \alpha_{ij}$ on the particular occasion with the particular individual. No constancy over repetitions is accorded to the e_{ij}.

To be poetic, μ_{ij} is an immutable constant of the Universe, α_{ij} is a lasting characteristic of the individual, and e_{ij} is but a fleeting aberration of the moment.

The terms μ_{ij} and α_{ij} in (3.1.1) are called respectively **fixed** and **random** effects; μ_{ij} is a fixed parameter of the set-up, taking a unique value irrespective of the individual, whereas α_{ij} varies randomly over the population of individuals. On the face of it, since μ_{ij} does not depend on the particular individual, the subscript i could be dropped and the fixed effect could be written as μ_j. This is all right when all the individuals in the sample are treated on an equal footing, with no differences in applied treatments or recorded backgrounds to distinguish them. But otherwise, e.g. when there are different treatment groups, it is convenient to retain the subscript i to indicate the special factors affecting the observations for that individual. The vector $\mu_i = (\mu_{i1}, \ldots, \mu_{ip})^{\mathrm{T}}$ corresponding to the p measurements on the ith individual will be referred to as the **μ-profile** for that individual.

To complete the specification of the model (3.1.1) some assumptions are required to formalize the above discussion. It is revealing to untangle these into three strata. The A-stratum below concerns just notation and terminology and so these are hardly 'assumptions' worthy of the name. The Bs specify interrelations between the random components of (3.1.1) in terms of their covariances, i.e. the correlation structure is laid down. In C the distributions are restricted to be normal. Most of the development proceeds quite happily under A and B; one only has to bring in C for the strict distributional results which support, e.g. the quoting of exact p-values for F-tests. E, V and C will denote respectively expectation, variance and covariance over hypothetical repetitions of the observations with individuals randomly selected from their population.

A1. The α_{ij}, for given j, vary over the population of individuals with mean 0 and variance $\sigma_{\alpha j}^2$: thus, $E(\alpha_{ij}) = 0$ and $V(\alpha_{ij}) = \sigma_{\alpha j}^2$. This is really no restriction at all since any nonzero mean is absorbed in μ_{ij}.

A2. The e_{ij}, for given j, vary over individuals with mean 0 and variance σ_j^2: thus, $E(e_{ij}) = 0$ and $V(e_{ij}) = \sigma_j^2$. As in A1, there is no real restriction here.

B1. For different i the α-profiles are uncorrelated, i.e. $C(\alpha_{ij}, \alpha_{i'j'}) = 0$ for $i \neq i'$. For given i, $C(\alpha_{ij}, \alpha_{ij'}) = \sigma_{\alpha jj'}$ (homogeneous over i but otherwise as yet unconstrained).

B2. The errors are all uncorrelated, i.e. $C(e_{ij}, e_{i'j'}) = 0$ if $i \neq i'$ or $j \neq j'$.

B3. The random components, α_{ij} and e_{ij}, are uncorrelated, i.e. $C(\alpha_{ij}, e_{i'j'}) = 0$ for all i, j, i', j'.

C. The α_{ij} and e_{ij} are normally distributed.

The consequences of assumptions A and B for the means and covariances of the y_{ij} are as follows. First, $E(y_{ij}) = \mu_{ij}$ follows simply from the zeroing of the means of α_{ij} and e_{ij} in A1 and A2. Second,

$$C(y_{ij}, y_{i'j'}) = E[(y_{ij} - \mu_{ij})(y_{i'j'} - \mu_{i'j'})] = E[(\alpha_{ij} + e_{ij})(\alpha_{i'j'} + e_{i'j'})]$$
$$= E[\alpha_{ij}\alpha_{i'j'} + \alpha_{ij}e_{i'j'} + \alpha_{i'j'}e_{ij} + e_{ij}e_{i'j'}].$$

Now, from A1 and B1, $E(\alpha_{ij}\alpha_{i'j'}) = \delta_{ii'}\sigma_{\alpha jj'}$ where $\delta_{ii'}$ is the Kronecker delta, taking values 1 when $i = i'$ and 0 when $i \neq i'$. From A2 and

B2, $E(e_{ij}e_{i'j'}) = \delta_{ii'}\delta_{jj'}\sigma_j^2$. From B3, $E(\alpha_{ij}e_{i'j'}) = 0$ and $E(\alpha_{i'j'}e_{ij}) = 0$. Thus the model specification for the first two moments of the data is

$$E(y_{ij}) = \mu_{ij}, \quad C(y_{ij}, y_{i'j'}) = \delta_{ii'}(\sigma_{\alpha jj'} + \delta_{jj'}\sigma_j^2). \tag{3.1.2}$$

While observations from different individuals $(i \neq i')$ are uncorrelated, according to (3.1.2), those on the same individual have correlations $\sigma_{\alpha jj'}/\{(\sigma_{\alpha jj} + \sigma_j^2)(\sigma_{\alpha j'j'} + \sigma_{j'}^2)\}^{1/2}$. In the special but commonly applied case where $\sigma_j^2 = \sigma^2$ and $\sigma_{\alpha jj'} = \sigma_\alpha^2$ (for each j, j') this is $\sigma_\alpha^2/(\sigma_\alpha^2 + \sigma^2)$. This correlation ranges from 0 to 1 as σ_α^2/σ^2 ranges from 0 to ∞. It measures the strength of the 'personal touch' and, as a correlation between different measurements on the same individual, is called an **intraclass correlation**. Notice how, starting with basically independent observations, correlations arise naturally through the random effects α_{ij}. More complex situations give rise to different specifications for $\sigma_{\alpha jj'}$ (e.g. Monlezun, Blouin and Malone, 1984; Goldstein, 1987, Chapter 2).

The above account gives a general set-up for so-called **mixed models** in which some effects are fixed (μ_{ij}) and some are random (α_{ij}). To see how (3.1.2) looks in practice some examples will now be tackled. In line with standard practice it will be assumed that σ_j^2 is constant over j and that $\sigma_{\alpha jj'}$ takes one value for $j = j'$ and a possibly different value for $j \neq j'$. The more general treatment, where these simplifying assumptions are not made, will be covered in the next chapter.

Example 3.1

The effect of a vitamin E diet supplement on the growth of guinea pigs was investigated as follows. For each animal the body weight was recorded at the end of weeks 1, 3, 4, 5, 6 and 7. All animals were given a growth-inhibiting substance during week 1, and the vitamin E therapy was started at the beginning of week 5. Three groups of animals, numbering five in each, received respectively zero, low and high doses of vitamin E. The body weights (in grams) are given in Table 3.1 and are plotted in Fig. 3.1 where much variation is evident.

The main question concerns possible differences in the growth profiles between groups and this will be tackled in the next section. For now, we are only interested in the distributional structure of the data.

Table 3.1. *Effect of diet supplement on growth rates*

Animal		Weeks 1	3	4	5	6	7
Group 1	1	455	460	510	504	436	466
	1	467	565	610	596	542	587
	3	445	530	580	597	582	619
	4	485	542	594	583	611	612
	5	480	500	550	528	562	576
Group 2	6	514	560	565	524	552	597
	7	440	480	536	484	567	569
	8	495	570	569	585	576	677
	9	520	590	610	637	671	702
	10	503	555	591	605	649	675
Group 3	11	496	560	622	622	632	670
	12	498	540	589	557	568	609
	13	478	510	568	555	576	605
	14	545	565	580	601	633	649
	15	472	498	540	524	532	583

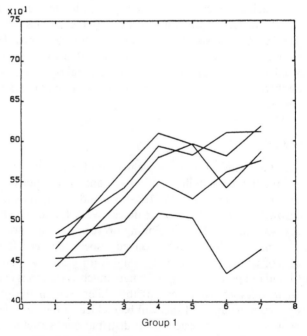

Fig. 3.1. *Effect of diet supplement on growth rates.*

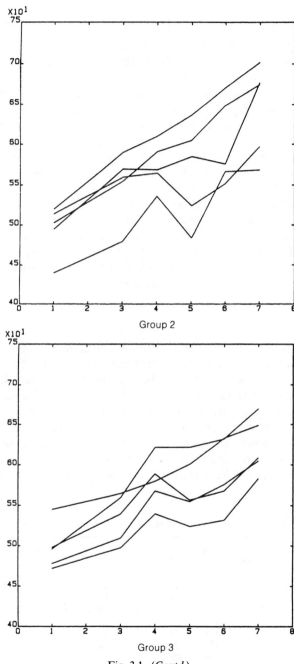

Group 2

Group 3

Fig. 3.1. (*Contd.*)

Suppose first that the random effect α_{ij} involves only overall differences in size between animals, there being no differences in the shape of the growth profile between animals in the same treatment group. In Anova terms this assumes that there is no interaction between individuals and measurement occasions. Thus $\alpha_{ij} = \alpha_i^I$, representing the adjustment to μ_{ij} for the overall level of y in individual i. In (3.1.2) we then have $\sigma_{\alpha jj'} = C(\alpha_{ij}, \alpha_{ij'}) = V\alpha_i^I = \sigma_I^2$, representing variation between individuals, i.e. between animal sizes.

Now suppose that we do allow for random variation of growth patterns between animals in the model. Thus $\alpha_{ij} = \alpha_i^I + \alpha_{ij}^{IO}, \alpha_{ij}^{IO}$ being the random interaction effect '*individuals* × *occasions*'. There is some discussion in section 3.3 which results in the assumption that α_i^I and α_{ij}^{IO} are uncorrelated, and that α_{ij}^{IO} and $\alpha_{ij'}^{IO}$ have correlation $-1/(p-1)$ for $j \neq j'$. Hence $\sigma_{\alpha jj'} = C(\alpha_i^I + \alpha_{ij}^{IO}, \alpha_i^I + \alpha_{ij'}^{IO}) = \sigma_I^2 + \sigma_{IO}^2$ for $j = j'$, $\sigma_{\alpha jj'} = \sigma_I^2 - \sigma_{IO}^2/(p-1)$ for $j \neq j'$, where $\sigma_{IO}^2 = V(\alpha_{ij}^{IO})$. The quantities σ_I^2 and σ_{IO}^2 are known as **variance components**. □

Example 3.2

Visual acuity was the subject of this investigation in which response times of the eyes to a stimulus were measured. The recorded variable was the time lag (milliseconds) between the stimulus (a light flash) and the electrical response at the back of the cortex. Seven student volunteers came forth and saw the light, recordings being made for left and right eyes each through lenses of powers 6/6, 6/18, 6/36 and 6/60. (The power 6/18 indicates that the magnification is such that the eye will perceive as being at 6 feet an object actually positioned at a distance of 18 feet.) The question is whether response time varies

Table 3.2. *Visual acuity with varying lens strength*

Subject	Left eye				Right eye			
	6/6	6/18	6/36	6/60	6/6	6/18	6/36	6/60
1	116	119	116	124	120	117	114	122
2	110	110	114	115	106	112	110	110
3	117	118	120	120	120	120	120	124
4	112	116	115	113	115	116	116	119
5	113	114	114	118	114	117	116	112
6	119	115	94	116	100	99	94	97
7	110	110	105	118	105	105	115	115

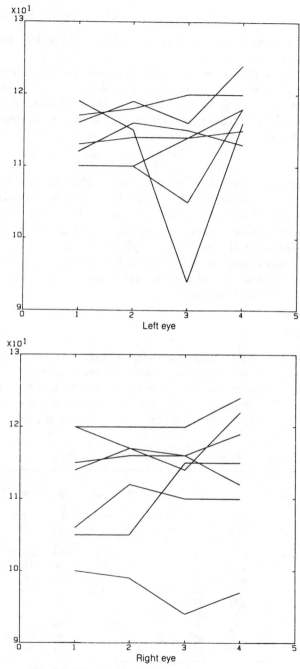

Fig. 3.2. *Visual acuity with varying lens strength.*

with lens strength. Table 3.2 gives the figures, and Fig. 3.2 shows the not very clear-cut trends.

There are $n = 7$ individuals and $p = 8$ measurements here. However, the measurements form a two-way layout with factor A (eyes) at two levels and factor B (lenses) at four levels. Thus, a full model for the individual random effects is (writing α_{ijk} rather than α_{ij} for notational convenience)

$$\alpha_{ijk} = \alpha_i^I + \alpha_{ij}^{IA} + \alpha_{ik}^{IB} + \alpha_{ijk}^{IAB},$$

representing contributions from I (individual levels), I × A interaction, I × B interaction and I × A × B interaction. According to assumptions detailed in section 3.3 these four random effects are uncorrelated with variances $\sigma_I^2, \sigma_{IA}^2, \sigma_{IB}^2$ and σ_{IAB}^2. The covariance structure can be derived with some effort (section 3.5).

The analysis given later for this example will centre on contrasts within individuals (i.e. between factors A and B) rather than across individuals as in Example 3.1. □

Example 3.3

Twelve hospital patients underwent a dietary regime treatment. Observations were made on nine variables on each of seven occasions: twice before, thrice during, and twice after the treatment regime. The timing of the seven occasions was at week numbers 1, 2, 6, 10, 14, 15, 16. The record for variable 1 (plasma ascorbic acid)

Table 3.3. *Reaction of twelve patients to treatment*

Patient	Week 1	2	6	10	14	15	16
1	0.22	0.00	1.03	0.67	0.75	0.65	0.59
2	0.18	0.00	0.96	0.96	0.98	1.03	0.70
3	0.73	0.37	1.18	0.76	1.07	0.80	1.10
4	0.30	0.25	0.74	1.10	1.48	0.39	0.36
5	0.54	0.42	1.33	1.32	1.30	0.74	0.56
6	0.16	0.30	1.27	1.06	1.39	0.63	0.40
7	0.30	1.09	1.17	0.90	1.17	0.75	0.88
8	0.70	1.30	1.80	1.80	1.60	1.23	0.41
9	0.31	0.54	1.24	0.56	0.77	0.28	0.40
10	1.40	1.40	1.64	1.28	1.12	0.66	0.77
11	0.60	0.80	1.02	1.28	1.16	1.01	0.67
12	0.73	0.50	1.08	1.26	1.17	0.91	0.87

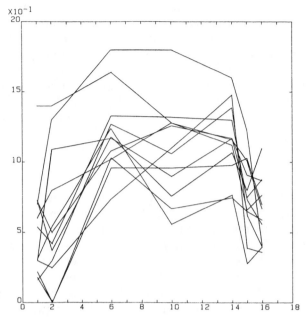

Fig. 3.3. *Reaction of twelve patients to treatment.*

appears in Table 3.3, and the corresponding picture is Fig. 3.3. The overall impression is of a sharp rise at week 6 and a fall at week 15, the whole being decorated by much variation both within and between patients. There are $n = 12$ individuals and $p = 7$ measurements. Unlike Example 3.1, it is the measurement occasions which are grouped here, into three phases (two pre-, three during, and two post-treatment), rather than the individuals. However, as in Example 3.1, an appropriate random effects model is $\alpha_{ij} = \alpha_i^I + \alpha_{ij}^{IO}$ and the covariance structure $\sigma_{\alpha jj'}$ follows as before. □

3.2 Anova

In the previous section some general structure was laid down for repeated measurements and some consequences were worked out for the covariances of the observations. Now we must get down to the more mundane business of analysing the data, trying to answer the investigators' questions. In the examples used in section 3.1 one would naturally think of Anova if each measurement came from a different individual, i.e. if all the observations were independent without the complication of being repeated. A traditional approach, which is the topic of this section, is to perform nevertheless a standard Anova and then to see what modifications, if any, are required to make the results valid. In the examples which follow the F-tests will be constructed from the Anova by common sense. Purists will find more formal justification below in section 3.5, and the not-so-pure will find a more practical discussion of the issues in section 3.6.

Example 3.1 (contd)

The growth profiles of the guinea pigs will be affected in some way by the treatment starting in week 5, but the first concern is simply to see whether profiles differ between the three groups.

The data form a two-way layout, *animals* ($n = 15$) × *weeks* ($p = 6$), and the usual table of sums of squares (s.s.) is constructed. Thus, in Table 3.4 the sum of squares between weeks (columns) is 142 554.50 and the s.s. between animals (rows) 123 982.26 has been partitioned into components between-groups and the residual between-animals-within-groups (a.w.g.). The mean-squares, not shown in Table 3.4, are obtained in the usual way as s.s./d.f., and the last column gives their expectations under the model described above. These expectations tell us what the various mean-squares are measuring and which variance ratios will be informative about the questions of interest. The horrendous detailed calculations of these expected

Table 3.4. *Anova for Example 3.1*

Source of variation	d.f.	s.s.	E(m.s.)
Weeks	5	142 554·50	$\sigma^2 + 6\sigma_{IO}^2/5 + D_O/5$
Animals	14	123 982·26	$\sigma^2 + 6\sigma_I^2 + D_G/14$
$\quad\{$ groups	$\{\ 2$	$\{\ 18\,548{\cdot}07$	$\{\sigma^2 + 6\sigma_I^2 + D_G/2$
$\quad\{$ a.w.g.	$\{\,12$	$\{\,105\,434{\cdot}19$	$\{\sigma^2 + 6\sigma_I^2$
Weeks × Animals	70	42 315·33	$\sigma^2 + 6\sigma_{IO}^2/5 + D_{GO}/70$
$\quad\{$ weeks × groups	$\{\,10$	$\{\ 9\,762{\cdot}73$	$\{\sigma^2 + 6\sigma_{IO}^2/5 + D_{GO}/10$
$\quad\{$ weeks × a.w.g.	$\{\,60$	$\{\,32\,552{\cdot}60$	$\{\sigma^2 + 6\sigma_{IO}^2/5$
Total	89	308 852·12	–

mean-squares are given in section 3.3 together with a government health warning. It suffices here to know that D_O, D_G and D_{GO} respectively measure differences between-occasions (weeks), between-groups, and the interaction *groups* × *occasions*. Similarly, σ^2, σ_I^2 and σ_{IO}^2 are the variances which measure respectively error variation, random variation between-individuals, and random *individuals* × *occasions* interaction.

The expected mean-squares tell us how to construct variance ratios. Thus, the null hypothesis $D_G = 0$ of zero group differences is tested by $F_{2,12} = (18\,548\cdot07/2) \div (105\,434\cdot19/12) = 1\cdot06$ ($p > 0\cdot10$). There is no other suitable pair of mean-squares, bearing in mind that s.s.(Animals) contains s.s.(a.w.g.) and is therefore not independent of it. Likewise, the interaction *weeks* × *groups* (null hypothesis $D_{GO} = 0$) is tested by $F_{10,60} = (9762\cdot73/10) \div (32\,552.60/60) = 1\cdot80$ ($p > 0\cdot05$). Thus, there seems to be no strong evidence for differences between groups in either overall profile level (quantified by D_G) or profile shape (D_{GO}); $D_{GO} = 0$ means that the mean profiles for different groups are parallel.

In this instance there is no question that there will be differences between weeks, since the animals are growing. However, for illustration only, the null hypothesis $D_O = 0$ (no differences between occasions) would be tested by $F_{5,60} = (142\,554\cdot50/5) \div (32\,552\cdot60/60) = 52.55$ ($p < 0\cdot001$). The justification for these F-tests is given below in section 3.5.

It is sometimes required to quote estimates of the variance components σ^2, σ_I^2 and σ_{IO}^2. Looking at Table 3.4 we see that only the combinations $\sigma^2 + 6\sigma_{IO}^2/5$ and $\sigma^2 + 6\sigma_I^2$ occur in the E(m.s.) column, so only these are estimable. Thus $\hat{\sigma}^2 + 6\hat{\sigma}_I^2 = 105\,434\cdot19/12$ and $\hat{\sigma}^2 + 6\hat{\sigma}_{IO}^2/5 = 32\,552\cdot60/60$, from which $\hat{\sigma}_I^2 + \hat{\sigma}^2/6 = 1464\cdot3638$ and (by subtraction) $\hat{\sigma}_I^2 - \hat{\sigma}_{IO}^2/5 = 1373\cdot9399$. It follows that $1373\cdot9399 < \hat{\sigma}_I^2 < 1464\cdot3638$ and that $\max(\hat{\sigma}^2, \hat{\sigma}^2 + 6\hat{\sigma}_{IO}^2/5) < 6(1464\cdot3638 - 1373\cdot9399) = 542\cdot5434$. More formal estimation will be covered in Chapter 5. □

Example 3.2 (contd)

The Anova here is carried out as for a straight three-way factorial with *subjects*($n = 7$) × *eyes*(left, right) × *lenses*(6, 18, 36, 60). Table 3.5 gives the sums of squares, etc., the E(m.s.) forms being calculated in section 3.3. The differences between eyes, between lenses, and their

Table 3.5. *Anova for Example 3.2*

Source of variation	d.f.	s.s.	E(m.s.)
Subjects	6	1379·429	$\sigma^2 + 8\sigma_I^2$
Eyes	1	46·446	$\sigma^2 + 8\sigma_{IA}^2 + D_A$
Lenses	3	140·768	$\sigma^2 + 8\sigma_{IB}^2/3 + D_B/3$
Subjects × Eyes	6	357·429	$\sigma^2 + 8\sigma_{IA}^2$
Subjects × Lenses	18	375·857	$\sigma^2 + 8\sigma_{IB}^2/3$
Eyes × Lenses	3	40·625	$\sigma^2 + 8\sigma_{IAB}^2/3 + D_{AB}/3$
Residual	18	231·000	$\sigma^2 + 8\sigma_{IAB}^2/3$
Total	55	2571·554	–

interaction are quantified by D_A, D_B, and D_{AB} respectively. The variance components $\sigma^2, \sigma_I^2, \sigma_{IA}^2, \sigma_{IB}^2$ and σ_{IAB}^2 correspond respectively to error, I (individual variation) and random interactions $I \times A$, $I \times B$, $I \times A \times B$.

The F-tests for the factors of interest are determined by inspection of the expected mean-squares. They are as follows:

1. Eyes: $F_{1,6} = (46 \cdot 446/1) \div (357 \cdot 429/6) = 0 \cdot 78$ ($p > 0 \cdot 10$).
2. Lenses: $F_{3,18} = (140 \cdot 768/3) \div (375 \cdot 857/18) = 2.25$ ($p > 0 \cdot 10$).
3. Eyes × Lenses: $F_{3,18} = (40 \cdot 625/3) \div (231 \cdot 000/18) = 1 \cdot 06$ ($p > 0 \cdot 10$).

The news will have to be broken to the experimenter that nothing shows up significant, thus robbing his report of the inevitable line 'the eyes have it'.

Estimation of the variance components rests on the four mean-squares for *subjects*, *subjects × eyes*, *subjects × lenses* and *residual*, these being the ones corresponding to random effects. Unfortunately, there are five unknowns: $\sigma^2, \sigma_I^2, \sigma_{IA}^2, \sigma_{IB}^2$ and σ_{IAB}^2. However, as in Example 3.1, we can make some relative estimates. For instance, $\hat{\sigma}^2 + 8\hat{\sigma}_I^2 = 1379 \cdot 429/6$ and $\hat{\sigma}^2 + 8\hat{\sigma}_{IAB}^2/3 = 231 \cdot 000/18$, leading to $\hat{\sigma}_I^2 + \hat{\sigma}^2/8 = 28 \cdot 74$ and $\hat{\sigma}_I^2 - 8\hat{\sigma}_{IAB}^2/3 = 27 \cdot 13$, thence to $27 \cdot 13 < \hat{\sigma}_I^2 < 28 \cdot 74$. □

Example 3.3 (contd)

The Anova here is like that for Example 3.1 but now it is the columns (*occasions*) which are grouped rather than the rows (*individuals*).

Table 3.6. *Anova for Example 3.3*

Source of variation	d.f.	s.s.	E(m.s.)
Patients	11	3·6820	$\sigma^2 + 7$
Occasions	6	6·1645	$\sigma^2 + 7\sigma_{IO}^2/6 + D_O/6$
$\begin{cases}\text{Treatment phases} \\ \text{occ. within phases}\end{cases}$	$\begin{cases}2 \\ 4\end{cases}$	$\begin{cases}5\cdot9611 \\ 0\cdot2034\end{cases}$	$\begin{cases}\sigma^2 + 7\sigma_{IO}^2/2 + D_O/2 \\ \sigma^2\end{cases}$
Patients × Occasions	66	4·4491	$\sigma^2 + 7\sigma_{IO}^2/6$
$\begin{cases}\text{Patients} \times \text{phases} \\ \text{Patients} \times \text{o.w.p.}\end{cases}$	$\begin{cases}22 \\ 44\end{cases}$	$\begin{cases}2\cdot3969 \\ 2\cdot0522\end{cases}$	$\begin{cases}\sigma^2 + 7\sigma_{IO}^2/2 \\ \sigma^2\end{cases}$
Total	83	14·2957	–

Table 3.6 gives the figures together with the expected mean-squares from section 3.3; the various symbols have the (by now) usual meaning (o.w.p. = occasions within phases).

There is only one F-test of interest, this being in connection with the null hypothesis $D_O = 0$ of no differences between phases. (Recall that the seven occasions are grouped into three phases, these being two pre-treatment, three during, and two post-treatment.) Thus, after inspecting the E(m.s.) column to define a suitable ratio, $F_{2,22} = (5\cdot9611/2) \div (2\cdot3969/22) = 27\cdot36$ ($p < 0\cdot001$).

We can estimate the variance components as follows. From the *patients × o.w.p.* line $\hat{\sigma}^2 = 2\cdot0522/44 = 0.047$; alternatively, from the *occasions-within-phases* line $\hat{\sigma}^2 = 0\cdot2034/4 = 0\cdot051$ but this is based on far fewer degrees of freedom. From s.s.(Patients) we have $\hat{\sigma}^2 + 7\hat{\sigma}_I^2 = 3\cdot6820/11 = 0\cdot033$, thus $\hat{\sigma}_I^2 = (0\cdot033 - 0\cdot047)/7 = -0\cdot002$. This is embarrassing – we have a negative estimate for a positive quantity. The simplest dodge is to decide that σ_I^2 must be 'small' and perhaps report an estimate of zero, thus avoiding derision from the client. However, this destroys the formal property of unbiasedness of these Anova-method estimators. The point is discussed at length by Searle (1971, section 9.8b). □

3.3 Calculation of expected mean-squares

The algebraic forms of the E(m.s.) terms in the Anova tables are derived here in detail. They result simply from applying (3.1.2) to the particular set-up under study. Only layouts of the general forms of the examples used in section 3.2 will be treated. There are so many variations on even the simplest designs that comprehensive treatment is not possible. However, it is hoped that the material here will be sufficiently illustrative to show what is involved generally, and also to enable the reader to decide whether he wants to be involved generally. The method employed is the most direct attack based on the simple identity $E(y^2) = \{E(y)\}^2 + V(y)$. In the following section a general, automated method will be described based on matrix manipulation.

On notation, wherever a subscript is replaced by $+$ it denotes summation over that index. Thus, for example, y_{i+} means $y_{i1} + y_{i2} + \cdots + y_{ip}$ and μ_{++} means $\sum_i \sum_j \mu_{ij}$. The errors e_{ij} will have variance $V(e_{ij}) = \sigma^2$ throughout, assumed independent of j.

A common prototype model for the random effects α_{ij} in (3.1.1) is $\alpha_{ij} = \alpha_i^I + \alpha_{ij}^{IO}$. Here α_i^I is the individual mean level with variance $V(\alpha_i^I) = \sigma_I^2$ over the population. Also, α_{ij}^{IO} is the random *individuals × occasions* interaction with variance $V(\alpha_{ij}^{IO}) = \sigma_{IO}^2$ (homogeneous over j) and satisfying the usual identifiability constraint $\alpha_{i+}^{IO} = 0$. Three remarks are in order:

1. The other constraint, $\alpha_{+j}^{IO} = 0$, which would normally be made for fixed effects, is not sensible here. This is because it would presuppose that the random selection of individuals for the sample always came up with a set of precisely related α_{ij}^{IO} values, i.e. related by $\alpha_{+j}^{IO} = 0$.

2. The constraint that is made, $\alpha_{i+}^{IO} = 0$, implies that the α_{ij}^{IO} are correlated as follows:

$$0 = V(\alpha_{i+}^{IO}) = \sum_{jj'} C(\alpha_{ij}^{IO}, \alpha_{ij'}^{IO}) = p\sigma_{IO}^2 + p(p-1)\sigma_{IO}^2\rho$$

   where ρ denotes the correlation between α_{ij}^{IO} and $\alpha_{ij'}^{IO}$, assumed to be the same for each pair $j \neq j'$. Hence $\rho = -1/(p-1)$.

3. We will follow Searle (1971) and Winer (1971) in assuming that α_i^I and α_{ij}^{IO} are uncorrelated. Scheffé (1959, section 8.1) gives a more careful analysis in which much weaker assumptions are made but, because of the way the algebra works out, comes up with expected mean-squares of the same form for the two-way mixed factorial layout.

Example 3.1 (contd)

The general layout here is n *individuals × p occasions* with the individuals divided into G groups of sizes n_g $(g = 1, \ldots, G)$. In the general model (3.1.1) we take $\mu_{ij} = \mu_j^{(g)}$ for $i \in g$; '$i \in g$' will be used as an abbreviation for 'individual i belonging to group g'. Also, the random effects are modelled as $\alpha_{ij} = \alpha_i^I + \alpha_{ij}^{IO}$. Hence, in (3.1.2), $\sigma_{\alpha jj'} = \sigma_I^2 + \sigma_{IO}^2$ for $j = j'$ and $\sigma_{\alpha jj'} = \sigma_I^2 - \sigma_{IO}^2/(p-1)$ for $j \neq j'$.

1. The total sum of squares is $\mathrm{ssTot} = \sum_{ij} y_{ij}^2 - y_{++}^2/np$. Now,

$$E(y_{ij}) = \mu_{ij}, \quad V(y_{ij}) = \sigma_I^2 + \sigma_{IO}^2 + \sigma^2, \quad E(y_{++}) = \mu_{++},$$
$$V(y_{++}) = nVy_{1+} = nV(p\alpha_1^I + \alpha_{1+}^{IO} + e_{1+}) = n(p^2\sigma_I^2 + p\sigma^2).$$

Hence, using the basic identity $E(y^2) = \{E(y)\}^2 + V(y)$, we find that

$$E(\mathrm{ssTot}) = D_{\mathrm{Tot}} + (n-1)p\sigma_I^2 + np\sigma_{IO}^2 + (np-1)\sigma^2$$

where $D_{\mathrm{Tot}} = \sum_{ij}\mu_{ij}^2 - \mu_{++}^2/np = \sum_{ij}(\mu_{ij} - \mu_{++}/np)^2$ represents differences among the μ_{ij}.

2. The sum of squares between individuals is $\mathrm{ssI} = \sum_i y_{i+}^2/p - y_{++}^2/np$. Thus,

$$E(y_{i+}) = \mu_{i+}, \quad V(y_{i+}) = V(p\alpha_i^I + \alpha_{i+}^{IO} + e_{i+}) = p^2\sigma_I^2 + p\sigma^2,$$
$$E(\mathrm{ssI}) = D_I + (n-1)p\sigma_I^2 + (n-1)\sigma^2$$

where $D_I = \sum_i \mu_{i+}^2/p - \mu_{++}^2/np = p^{-1}\sum_i(\mu_{i+} - \mu_{++}/n)^2$ represents differences among the μ_{i+}.

3. The sum of squares between occasions is $\mathrm{ssO} = \sum_j y_{+j}^2/n - y_{++}^2/np$. Thus,

$$E(y_{+j}) = \mu_{+j}, \quad V(y_{+j}) = nV(y_{1j}) = nV(\alpha_1^I + \alpha_{1j}^{IO} + e_{1j})$$
$$= n(\sigma_I^2 + \sigma_{IO}^2 + \sigma^2),$$
$$E(\mathrm{ssO}) = D_O + p\sigma_{IO}^2 + (p-1)\sigma^2$$

where $D_O = \sum_j \mu_{+j}^2/n - \mu_{++}^2/np = n^{-1}\sum_j(\mu_{+j} - \mu_{++}/p)^2$ represents differences among the μ_{+j}.

4. The sum of squares between groups is $\mathrm{ssG} = \sum_g (\sum_{i \in g} y_{i+})^2/n_g p - y_{++}^2/np$. Thus,

$$E\left(\sum_{i \in g} y_{i+}\right) = n_g\mu_+^{(g)}, \quad V\left(\sum_{i \in g} y_{i+}\right) = n_g V(y_{i+}) = n_g(p^2\sigma_I^2 + p\sigma^2),$$
$$E(\mathrm{ssG}) = D_G + (G-1)p\sigma_I^2 + (G-1)\sigma^2$$

where $D_G = p^{-1}\sum_g n_g\mu_+^{(g)2} - \mu_{++}^2/np = p^{-1}\sum_g n_g(\mu_+^{(g)} - \mu_{++}/n)^2$ represents differences among the $\mu_+^{(g)}$. It is easy to show that $D_I = D_G$, which means that the μ-differences between individuals are just group differences, this being a consequence of the μ-model.

5. The sum of squares for interaction *groups* × *occasions* is ssGO = $\sum_g \sum_j (\sum_{i \in g} y_{ij})^2 / n_g - y_{++}^2/np - \text{ssG} - \text{ssO}$. Thus

$$E\left(\sum_{i \in g} y_{ij}\right) = n_g \mu_j^{(g)}, \quad V\left(\sum_{i \in g} y_{ij}\right) = n_g V(y_{ij}) = n_g(\sigma_1^2 + \sigma_{IO}^2 + \sigma^2),$$

$$E(\text{ssGO}) = D_{GO} + (G-1)p\sigma_{IO}^2 + (G-1)(p-1)\sigma^2$$

where $D_{GO} = \sum_g \sum_j n_g (\mu_j^{(g)} - \mu_{+j}/n - \mu_+^{(g)}/p + \mu_{++}/np)^2$ represents differences between groups in the μ-profile. D_{GO} has the familiar form of an interaction s.s. for the μ_{ij} and it is easy to show that $D_{Tot} = D_O + D_G + D_{GO}$.

6. The remaining sums of squares, and their expectations, can be found by subtraction. Thus, the s.s. for interaction *individuals* × *occasions* is ssIO = ssTot − ssI − ssO, with expectation $D_{GO} + (n-1)p\sigma_{IO}^2 + (n-1)(p-1)\sigma^2$, the s.s. for *individuals-within-groups* is ssI − ssG, with expectation $(n-G)p\sigma_I^2 + (n-G)\sigma^2$, and the s.s. for interaction *individuals-within-groups* × *occasions* is ssIO − ssGO, with expectation $(n-G)p\sigma_{IO}^2 + (n-G)(p-1)\sigma^2$. □

Example 3.2 (contd)

Each of the n individuals is observed over a complete $a \times b$ factorial design with factors A and B, say. Thus $p = ab$, and the data y_{ij} can be alternatively represented as Y_{ijk} (ith individual, jth level of A, kth level of B) for convenience in writing down the various sums of squares in familiar forms. In the model (3.1.1) the μ-term will then be written as μ_{jk}, the same for all i. In more extended form μ_{jk} might be expressed in the usual way as $\mu + \mu_j^A + \mu_k^B + \mu_{jk}^{AB}$ with the usual identifiability constraints. The α-term will be taken as $\alpha_i^I + \alpha_{ij}^{IA} + \alpha_{ik}^{IB} + \alpha_{ijk}^{IAB}$, these random effects being uncorrelated with variances σ_I^2, σ_{IA}^2, σ_{IB}^2 and σ_{IAB}^2. Also, constraints $\alpha_{i+}^{IA} = \alpha_{i+}^{IB} = \alpha_{i+k}^{IAB} = \alpha_{ij+}^{IAB} = 0$ will hold, implying certain correlations like that discussed for α_{ij}^{IO} in Example 3.1.

1. Total s.s: ssTot = $\sum_{ijk} Y_{ijk}^2 - Y_{+++}^2/nab$. Hence,

$$E(Y_{ijk}) = \mu_{jk}, \quad V(Y_{ijk}) = \sigma_I^2 + \sigma_{IA}^2 + \sigma_{IB}^2 + \sigma_{IAB}^2 + \sigma^2,$$

$$E(Y_{+++}) = n\mu_{+++},$$

$$V(Y_{+++}) = nV(Y_{1++}) = nV(ab\alpha_1^I + b\alpha_{1+}^{IA} + a\alpha_{1+}^{IB} + \alpha_{1++}^{IAB} + e_{1++})$$
$$= n(a^2b^2\sigma_I^2 + ab\sigma^2),$$

$$E(\text{ssTot}) = D_{Tot} + (n-1)ab\sigma_I^2 + nab(\sigma_{IA}^2 + \sigma_{IB}^2 + \sigma_{IAB}^2) + (nab-1)\sigma^2$$

where $D_{Tot} = n\sum_{jk} \mu_{jk}^2 - n\mu_{++}^2/ab$.

2. Individuals: $\text{ssI} = \sum_i Y_{i++}^2/ab - Y_{+++}^2/nab$. Hence,

$$E(Y_{i++}) = \mu_{++}, \quad V(Y_{i++}) = ab(ab\sigma_I^2 + \sigma^2),$$
$$E(\text{ssI}) = (n-1)ab\sigma_I^2 + (n-1)\sigma^2.$$

3. Factors A and B: $\text{ssA} = \sum_j Y_{+j+}^2/nb - Y_{+++}^2/nab$. Now, $E(Y_{+j+}) = n\mu_{j+}$ and

$$V(Y_{+j+}) = nV(b\alpha_1^I + b\alpha_{1j}^{IA} + \alpha_{1+}^{IB} + \alpha_{1j+}^{IAB} + e_{1j+})$$
$$= n(b^2\sigma_I^2 + b^2\sigma_{IA}^2 + b\sigma^2)$$

so

$$E(\text{ssA}) = D_A + ab\sigma_{IA}^2 + (a-1)\sigma^2$$

where $D_A = n\sum_j \mu_{j+}^2/b - n\mu_{++}^2/ab$. Similarly,

$$E(\text{ssB}) = D_B + ab\sigma_{IB}^2 + (b-1)\sigma^2$$

where $D_B = n\sum_k \mu_{+k2}/a - n\mu_{++}^2/ab$. The interaction AB has

$$\text{ssAB} = \sum_{jk} Y_{+jk}^2/n - Y_{+++}^2/nab - \text{ssA} - \text{ssB}.$$

Now, $E(Y_{+jk}) = n\mu_{jk}$ and

$$V(Y_{+jk}) = nV(\alpha_1^I + \alpha_{1j}^{IA} + \alpha_{1k}^{IB} + \alpha_{1jk}^{IAB} + e_{1jk})$$
$$= n(\sigma_I^2 + \sigma_{IA}^2 + \sigma_{IB}^2 + \sigma_{IAB}^2 + \sigma^2)$$

so $E(\text{ssAB}) = D_{AB} + ab\sigma_{IAB}^2 + (a-1)(b-1)\sigma^2$ where

$$D_{AB} = D_{Tot} - D_A - D_B = n\sum_{jk}(\mu_{jk} - \mu_{j+}/b - \mu_{+k}/a + \mu_{++}/ab)^2.$$

4. Interactions $I \times A$ and $I \times B$: $\text{ssIA} = \sum_{ij} Y_{ij+}^2/b - Y_{+++}^2/nab - \text{ssI} - \text{ssA}$. Hence,

$$E(Y_{ij+}) = \mu_{j+}$$
$$V(Y_{ij+}) = V(b\alpha_i^I + b\alpha_{ij}^{IA} + \alpha_{i+}^{IB} + \alpha_{ij+}^{IAB} + e_{ij+}) = b^2\sigma_I^2 + b^2\sigma_{IA}^2 + b\sigma^2,$$
$$E(\text{ssIA}) = (n-1)ab\sigma_{IA}^2 + (n-1)(a-1)\sigma^2$$

and similarly $E(\text{ssIB}) = (n-1)ab\sigma_{IB}^2 + (n-1)(b-1)\sigma^2$.

5. The residual s.s. for $I \times A \times B$ is found by subtraction so

$$E(\text{ssIAB}) = E(\text{ssTot} - \text{ssI} - \text{ssA} - \text{ssB} - \text{ssIA} - \text{ssIB} - \text{ssAB})$$
$$= (n-1)ab\sigma_{IAB}^2 + (n-1)(a-1)(b-1)\sigma^2. \quad \square$$

Example 3.3 (contd)

The layout is *n individuals* $\times$ *p occasions*, as in Example 3.1, but the occasions are grouped instead of the individuals. Let there be *H* sets of occasions, which will be dubbed 'phases', with p_h occasions in the hth phase, thus partitioning p into $p_1 + p_2 + \cdots + p_H = p$. Then, in the model (3.1.1) the μ-term is represented as $\mu_{ij} = \mu_h$ for $j \in h$; here '$j \in h$' means 'the jth occasion falling within the hth phase'. The α-term will satisfy the constraint $\sum_h p_h \alpha_{ih}^{IO} = 0$ to conform with $\alpha_i = p\alpha_i^I$.

The sums of squares, and their expectations, for Total (ssTot), *between-individuals* (ssI), *between-occasions* (ssO), and *individuals × occasions* (ssIO) are the same as in Example 3.1 with $G = 1$ (so D_I there is zero here).

1. *Phases*: $\mathrm{ssP} = \sum_h (\sum_{j \in h} y_{+j})^2 / n p_h - y_{++}^2 / np$. Hence,

$$E\left(\sum_{j \in h} y_{+j} \right) = n p_h \mu_h$$

$$V\left(\sum_{j \in h} y_{+j} \right) = nV\left(p_h \alpha_i^I + p_h \alpha_{ih}^{IO} + \sum_{j \in h} e_{ij} \right) = n(p_h^2 \sigma_I^2 + p_h^2 \sigma_{IO}^2 + p_h \sigma^2),$$

$$E(\mathrm{ssP}) = D_P + p \sigma_{IO}^2 + (H - 1)\sigma^2$$

where $D_P = n \sum_h p_h \mu_h^2 - \mu_{++}^2 / np$; D_P here is the D_O of Example 3.1.

2. *Individuals × Phases*: $\mathrm{ssIP} = \sum_i \sum_h (\sum_{j \in h} y_{ij})^2 / p_h - y_{++}^2 / np - \mathrm{ssI} - \mathrm{ssP}$. Hence,

$$E\left(\sum_{j \in h} y_{ij} \right) = p_h \mu_h, \quad V\left(\sum_{j \in h} y_{ij} \right) = p_h^2 \sigma_I^2 + p_h^2 \sigma_{IO}^2 + p_h \sigma^2,$$

$$E(\mathrm{ssIP}) = (n - 1) p \sigma_{IO}^2 + (n - 1)(H - 1)\sigma^2.$$

The sum of squares for *occasions-within-phases* is $\mathrm{ssO} - \mathrm{ssP}$, with expectation $(p - H)\sigma^2$, and that for the interaction *individuals × occasions-within-phases* is $\mathrm{ssIO} - \mathrm{ssIP}$, with expectation $(n - 1)(p - H)\sigma^2$. $\square$

3.4 Expected mean-squares by 'synthesis'

The examples treated in the previous section were fairly basic but nonetheless laborious. Hartley (1967) came up with a method for avoiding such 'algebraic heroics' by mechanizing the process, and this was generalized by Rao (1968). It works for any layout, balanced or not, and is particularly suited to routine computer processing – in fact, some statistical packages have it as an option. To follow the details in this section some familiarity with vector random variables and matrix manipulation is necessary.

As a preliminary we will illustrate the expression of Anova sums of squares and Anova models in matrix notation. Consider first a single sample $(y_1, \ldots, y_n)$ of n observations. The sum of squares (about the sample mean $\bar{y}$) is

$$\sum_i (y_i - \bar{y})^2 = \sum_i y_i^2 - (\sum_i y_i)^2/n = \mathbf{y}^T\mathbf{y} - (\mathbf{y}^T\mathbf{1}_n)^2/n$$

where $\mathbf{y}(n \times 1) = (y_1, \ldots, y_n)^T$ and $\mathbf{1}_n(n \times 1) = (1, \ldots, 1)^T$. Now $\mathbf{y}^T\mathbf{y} = \mathbf{y}^T\mathbf{I}_n\mathbf{y}$, where $\mathbf{I}_n(n \times n)$ is a unit matrix, and $(\mathbf{y}^T\mathbf{1}_n)^2 = \mathbf{y}^T\mathbf{J}_n\mathbf{y}$, where $\mathbf{J}_n(n \times n) = \mathbf{1}_n\mathbf{1}_n^T$ is a matrix of 1's. Hence the sum of squares is expressible as $\mathbf{y}^T\mathbf{I}_n\mathbf{y} - \mathbf{y}^T\mathbf{J}_n\mathbf{y}/n$, or $\mathbf{y}^T\mathbf{A}\mathbf{y}$ with $\mathbf{A} = \mathbf{I}_n - n^{-1}\mathbf{J}_n$. Note that trace $\mathbf{A} = n - 1$ here, the degrees of freedom. Now let us proceed to the one-way layout with y_{ij} representing the ith observation in the jth group $(i = 1, \ldots, n_j; j = 1, \ldots, G)$. The within- and between-groups sums of squares are respectively

$$\text{ssW} = \sum_j (n_j - 1)s_j^2, \quad \text{ssB} = \sum_j n_j(\bar{y}_j - \bar{y})^2$$

where $\bar{y}_j$ is the jth group mean, s_j^2 is the jth group variance, and $\bar{y}$ is the overall mean. From the single-sample case just dealt with $(n_j - 1)s_j^2 = \mathbf{y}_j^T\mathbf{A}_j\mathbf{y}_j$, where $\mathbf{y}_j(n_j \times 1)$ is the data vector for the jth group and $\mathbf{A}_j(n_j \times n_j) = \mathbf{I}_{n_j} - n_j^{-1}\mathbf{J}_{n_j}$. Hence $\text{ssW} = \mathbf{y}^T\mathbf{A}\mathbf{y}$ where $\mathbf{y}^T = (\mathbf{y}_1^T, \ldots, \mathbf{y}_G^T)$ is the total data vector of length $n_+ = \sum_j n_j$ and $\mathbf{A}(n_+ \times n_+) = \text{diag}(\mathbf{A}_1, \ldots, \mathbf{A}_G)$, a block-diagonal matrix. Likewise

$$\text{ssB} = \sum_j n_j\bar{y}_j^2 - n_+\bar{y}^2 = \sum_j n_j^{-1}(\mathbf{y}_j^T\mathbf{1}_{n_j})^2 - n_+^{-1}(\mathbf{y}^T\mathbf{1}_{n_+})^2$$
$$= \sum_j n_j^{-1}\mathbf{y}_j^T\mathbf{J}_{n_j}\mathbf{y}_j - n_+^{-1}\mathbf{y}^T\mathbf{J}_{n_+}\mathbf{y} = \mathbf{y}^T\mathbf{B}\mathbf{y}$$

where $\mathbf{B} = \text{diag}(n_1^{-1}\mathbf{J}_{n_1}, \ldots, n_G^{-1}\mathbf{J}_{n_G}) - n_+^{-1}\mathbf{J}_{n_+}$. Note that trace $\mathbf{A} = \sum_j \text{trace } \mathbf{A}_j = \sum_j(n_j - 1) = n_+ - G$ and trace $\mathbf{B} = \sum n_j^{-1}\text{ trace }\mathbf{J}_{n_j} - n_+^{-1}\text{ trace }\mathbf{J}_{n_+} = G - 1$, the respective degrees of freedom for ssW and ssB. For a final illustration consider Example 3.1 with notation as in section 3.3. The between-individuals sum of squares is

$$\text{ssI} = \sum_i y_{i+}^2/p - y_{++}^2/np = \sum_i \mathbf{y}_i^T\mathbf{J}_p\mathbf{y}_i/p - \mathbf{y}^T\mathbf{J}_{np}\mathbf{y}/np$$

where $\mathbf{y}_i(p \times 1)$ is the data vector for individual i and $\mathbf{y}^T = (\mathbf{y}_1^T, \ldots, \mathbf{y}_n^T)$ is the total data vector. Thus $\text{ssI} = \mathbf{y}^T\mathbf{A}\mathbf{y}$ with $\mathbf{A} = p^{-1}\text{ diag}(\mathbf{J}_p, \ldots, \mathbf{J}_p) - (np)^{-1}\mathbf{J}_{np}$ and trace $\mathbf{A} = n - 1$. Likewise, the between-groups sum of squares is

$$ssG = \sum_g \left(\sum_{i \in g} y_{i+} \right)^2 \bigg/ n_g p - y_{++}^2/np = \mathbf{y}^T\mathbf{B}\mathbf{y}/p - \mathbf{y}^T\mathbf{J}_{np}\mathbf{y}/np$$

where $\mathbf{B} = \text{diag}(n_1^{-1}\mathbf{J}_{n_1p},\ldots,n_G^{-1}\mathbf{J}_{n_Gp})$. Thus $ssG = \mathbf{y}^T\mathbf{C}\mathbf{y}$ with $\mathbf{C} = p^{-1}\mathbf{B} - (np)^{-1}\mathbf{J}_{np}$ and $\text{trace}\,\mathbf{C} = G - 1$. In consequence, the sum of squares for animals-within-groups is $ss(a.w.g) = ssI - ssG = \mathbf{y}^T(\mathbf{A} - \mathbf{C})\mathbf{y}$ with degrees of freedom $\text{trace}\,(\mathbf{A} - \mathbf{C}) = n - G$. The other sums of squares can be similarly glamourized in matrix notation.

The second half of the preliminaries concerns the expression of Anova models in matrix/vector form. Let us use Example 3.1 again for illustration. The model is $y_{ij} = \mu_{ij} + \alpha_{ij} + e_{ij}$ where $\mu_{ij} = \mu_j^{(g)}$ for the jth occasion in group g, and $\alpha_{ij} = \alpha_i^I + \alpha_{ij}^{IO}$ includes random individual levels α_i^I and random *individual × occasion* interactions α_{ij}^{IO}. In vector form $\mathbf{y}(np \times 1) = \boldsymbol{\mu} + \boldsymbol{\alpha} + \mathbf{e}$. Here $\boldsymbol{\mu} = \mathbf{X}\boldsymbol{\beta}$, where $\boldsymbol{\beta}^T = (\boldsymbol{\mu}_1^T,\ldots,\boldsymbol{\mu}_G^T)$ contains the group μ-profiles and $\mathbf{X} = \text{diag}(\mathbf{X}_1,\ldots,\mathbf{X}_G)$ with $\mathbf{X}_j^T = (\mathbf{I}_p,\ldots,\mathbf{I}_p)$. Also $\boldsymbol{\alpha} = \mathbf{Z}_1\mathbf{b}_1 + \mathbf{Z}_2\mathbf{b}_2$, where $\mathbf{Z}_1 = \text{diag}(\mathbf{1}_p,\ldots,\mathbf{1}_p)$, $\mathbf{b}_1^T = (\alpha_1^I,\ldots,\alpha_n^I)$, $\mathbf{Z}_2 = \mathbf{I}_{np}$, and $\mathbf{b}_2^T = ((\boldsymbol{\alpha}_1^{IO})^T,\ldots,(\boldsymbol{\alpha}_n^{IO})^T)$ with $(\boldsymbol{\alpha}_i^{IO})^T = (\alpha_{i1}^{IO},\ldots,\alpha_{ip}^{IO})$. Thus the model is $\mathbf{y} = \mathbf{X}\boldsymbol{\beta} + \sum \mathbf{Z}_j\mathbf{b}_j + \mathbf{e}$ in illustration of (3.4.2) below.

Let $Q(\mathbf{y}) = \mathbf{y}^T\mathbf{A}\mathbf{y}$ be a mean-square in the Anova of the observation vector $\mathbf{y}$ with matrix $\mathbf{A}$, and let $\boldsymbol{\mu} = E(\mathbf{y})$ be the mean vector and $\boldsymbol{\Sigma} = V(\mathbf{y})$ be the covariance matrix of $\mathbf{y}$. Then

$$\begin{aligned}
E[Q(\mathbf{y})] &= E[(\mathbf{y} - \boldsymbol{\mu})^T\mathbf{A}(\mathbf{y} - \boldsymbol{\mu}) + 2\mathbf{y}^T\mathbf{A}\boldsymbol{\mu} - \boldsymbol{\mu}^T\mathbf{A}\boldsymbol{\mu}] \\
&= E[\text{trace}\,\{(\mathbf{y} - \boldsymbol{\mu})(\mathbf{y} - \boldsymbol{\mu})^T\mathbf{A}\}] + \boldsymbol{\mu}^T\mathbf{A}\boldsymbol{\mu} \\
&= \text{trace}\,(\boldsymbol{\Sigma}\mathbf{A}) + \boldsymbol{\mu}^T\mathbf{A}\boldsymbol{\mu}
\end{aligned} \tag{3.4.1}$$

Now suppose that $\mathbf{y}$ is subject to an Anova (or regression) model of form

$$\mathbf{y} = \mathbf{X}\boldsymbol{\beta} + \sum_j \mathbf{Z}_j\mathbf{b}_j + \mathbf{e} \tag{3.4.2}$$

where $\mathbf{X}$ is the design matrix for the effects $\boldsymbol{\beta}$, $\mathbf{e}$ is the 'error', and the $\mathbf{Z}_j$ are the design matrices for the random effects $\mathbf{b}_j$. The correspondence with the fundamental model (3.1.1) is clear. Denote the covariance matrix of $\mathbf{e}$ by $\mathbf{E}$, that of $\mathbf{b}_j$ by $\mathbf{B}_j$, and assume that the $\mathbf{b}_j$'s and $\mathbf{e}$ are otherwise uncorrelated, and that all have mean 0. In fact, $\sum \mathbf{Z}_j\mathbf{b}_j$ can be more tersely represented as $\mathbf{Z}\mathbf{b}$, with partitioning $\mathbf{Z} = (\mathbf{Z}_1,\mathbf{Z}_2,\ldots)$ and $\mathbf{b}^T = (\mathbf{b}_1^T,\mathbf{b}_2^T,\ldots)$, but we will stick with Rao's formulation for ease of reference. Thus, $\boldsymbol{\mu} = E(\mathbf{y}) = \mathbf{X}\boldsymbol{\beta}$

and

$$\Sigma = E[(\mathbf{y} - \boldsymbol{\mu})(\mathbf{y} - \boldsymbol{\mu})^{\mathrm{T}}] = \sum_j \mathbf{Z}_j \mathbf{B}_j \mathbf{Z}_j^{\mathrm{T}} + \mathbf{E}. \quad (3.4.3)$$

It follows from (3.4.1) that

$$E[Q(\mathbf{y})] = \boldsymbol{\beta}^{\mathrm{T}} \mathbf{X}^{\mathrm{T}} \mathbf{A} \mathbf{X} \boldsymbol{\beta} + \sum_j \operatorname{trace}(\mathbf{B}_j \mathbf{Z}_j^{\mathrm{T}} \mathbf{A} \mathbf{Z}_j) + \operatorname{trace}(\mathbf{E} \mathbf{A}). \quad (3.4.4)$$

The first case considered by Rao (1968, section 2) is where $\mathbf{E} = \sigma^2 \mathbf{I}_p$ and $\mathbf{B}_j = \sigma_j^2 \mathbf{I}_p$. The latter differs slightly from the structure assumed in section 3.3 here for the random effects in that the constraint (that the components of $\mathbf{b}_j$ sum to zero, and hence are correlated) is absent. Then $\operatorname{trace}(\mathbf{E}\mathbf{A}) = \sigma^2 \operatorname{trace} \mathbf{A}$ and, if $Q(\mathbf{y})$ is to represent a mean-square in the Anova, this must be just σ^2, so $\operatorname{trace} \mathbf{A} = 1$. Also, $\operatorname{trace}(\mathbf{B}_j \mathbf{Z}_j^{\mathrm{T}} \mathbf{A} \mathbf{Z}_j) = \sigma_j^2 \operatorname{trace}(\mathbf{Z}_j^{\mathrm{T}} \mathbf{A} \mathbf{Z}_j) = \sigma_j^2 \sum_r \mathbf{z}_{jr}^{\mathrm{T}} \mathbf{A} \mathbf{z}_{jr} = \sigma_j^2 \kappa_j$, say, where $\mathbf{z}_{jr}$ is the rth column of $\mathbf{Z}_j$. Now, κ_j is the coefficient of σ_j^2 in (3.4.4) and, in principle, can be calculated directly since $\mathbf{z}_{jr}$ is known and $\mathbf{A}$ can be found, though perhaps not easily. However, $\mathbf{z}_{jr}^{\mathrm{T}} \mathbf{A} \mathbf{z}_{jr} = Q(\mathbf{z}_{jr})$, the mean-square based on the synthetic data vector $\mathbf{z}_{jr}$. Thus, the numerical coefficient κ_j of σ_j^2 can be calculated simply by applying the Anova to each column $\mathbf{z}_{jr}$ of $\mathbf{Z}_j$ in turn, and adding the resulting $Q(\mathbf{z}_{jr})$. For the coefficients of the β-components in (3.4.4) we have

$$\boldsymbol{\beta}^{\mathrm{T}} \mathbf{X}^{\mathrm{T}} \mathbf{A} \mathbf{X} \boldsymbol{\beta} = \sum_{rs} \beta_r \beta_s (\mathbf{X}^{\mathrm{T}} \mathbf{A} \mathbf{X})_{rs} = \sum_{rs} \beta_r \beta_s (\mathbf{x}_r^{\mathrm{T}} \mathbf{A} \mathbf{x}_s)$$

where $\mathbf{x}_r$ is the rth column of $\mathbf{X}$. But

$$\mathbf{x}_r^{\mathrm{T}} \mathbf{A} \mathbf{x}_s = \tfrac{1}{2} \{ (\mathbf{x}_r + \mathbf{x}_s)^{\mathrm{T}} \mathbf{A} (\mathbf{x}_r + \mathbf{x}_s) - \mathbf{x}_r^{\mathrm{T}} \mathbf{A} \mathbf{x}_r - \mathbf{x}_s^{\mathrm{T}} \mathbf{A} \mathbf{x}_s \}$$
$$= \tfrac{1}{2} \{ Q(\mathbf{x}_r + \mathbf{x}_s) - Q(\mathbf{x}_r) - Q(\mathbf{x}_s) \} = \tau_{rs},$$

say. Hence, for the case considered, (3.4.4) yields

$$E[Q(\mathbf{y})] = \sum_{rs} \tau_{rs} \beta_r \beta_s + \sum_j \kappa_j \sigma_j^2 + \sigma^2 \quad (3.4.5)$$

where the coefficients κ_j and τ_{rs} can be computed automatically just by applying the Anova program to columns of $\mathbf{X}$ and $\mathbf{Z}$ posing as data vectors.

For the general case, where arbitrary structures occur for $\mathbf{E}$ and the $\mathbf{B}_j$, (3.4.4) may be evaluated similarly. Thus (Rao, 1968, section 4),

$$\operatorname{trace}(\mathbf{E} \mathbf{A}) = \sum_{rs} \mathbf{E}_{rs} \mathbf{A}_{rs}$$

where $\quad \mathbf{A}_{rs} = \mathbf{u}_r^{\mathrm{T}} \mathbf{A} \mathbf{u}_s = \tfrac{1}{2} \{ Q(\mathbf{u}_r + \mathbf{u}_s) - Q(\mathbf{u}_r) - Q(\mathbf{u}_s) \}, \quad \mathbf{u}_r \quad$ denoting

$(0, \ldots, 1, \ldots, 0)^{\mathrm{T}}$ with a 1 in the rth position and 0's elsewhere. Similarly,

$$\text{trace}(\mathbf{B}_j \mathbf{Z}_j^{\mathrm{T}} \mathbf{A} \mathbf{Z}_j) = \sum_{rs}(\mathbf{B}_j)_{rs}(\mathbf{Z}_j^{\mathrm{T}} \mathbf{A} \mathbf{Z}_j)_{rs} = \sum_{rs}(\mathbf{B}_j)_{rs}(\mathbf{z}_{jr}^{\mathrm{T}} \mathbf{A} \mathbf{z}_{js})$$

and $\mathbf{z}_{jr}^{\mathrm{T}} \mathbf{A} \mathbf{z}_{js} = \frac{1}{2}\{Q(\mathbf{z}_{jr} + \mathbf{z}_{js}) - Q(\mathbf{z}_{jr}) - Q(\mathbf{z}_{js})\} = \kappa_{jrs}$, say. Thus $E(Q(\mathbf{y}))$ is expressed in terms of the $\boldsymbol{\beta}_j$'s and the elements of the $\mathbf{B}_j$ and $\mathbf{E}$, with coefficients τ_{rs}, κ_{jrs} and $\mathbf{A}_{rs}$ computed numerically from the Anova program. Explicitly, in generalization of (3.4.5),

$$E[Q(\mathbf{y})] = \sum_{rs}\tau_{rs}\boldsymbol{\beta}_r\boldsymbol{\beta}_s + \sum_j\sum_{rs}\kappa_{jrs}(\mathbf{B}_j)_{rs} + \sum_{rs}\mathbf{A}_{rs}\mathbf{E}_{rs} \quad (3.4.6)$$

Rao (1968, section 3), following Hartley (1967, section 5), also extends the methodology to the computation of variances $V[Q(\mathbf{y})]$ and covariances $C[Q_1(\mathbf{y}), Q_2(\mathbf{y})]$ of mean-squares.

3.5 Contrasts, compound symmetry and F-tests

The purpose of this section is to provide theoretical justification for the F-tests used in the foregoing Anovas. A little familiarity with matrix and vector manipulation will be assumed. The special vectors $\mathbf{0}_p = (0, \ldots, 0)^{\mathrm{T}}$ and $\mathbf{1}_p = (1, \ldots, 1)^{\mathrm{T}}$, of length p, will be found useful.

Let us start with a familiar example, the two-sample t-test. Let $\mathbf{y} = (y_1, \ldots, y_r, y_{r+1}, \ldots, y_s)^{\mathrm{T}}$ be the vector of all $r + s$ independent observations, and let $\bar{y}_1$ and $\bar{y}_2$ be the two sample means. The t-value for testing the hypothesis H: $\mu_1 = \mu_2$, of equality of the underlying means, is just a scaled version of $\bar{y}_1 - \bar{y}_2$, and this difference can be written as $\mathbf{c}^{\mathrm{T}}\mathbf{y}$ with $\mathbf{c}^{\mathrm{T}} = (r^{-1}\mathbf{1}_r^{\mathrm{T}}, -s^{-1}\mathbf{1}_s^{\mathrm{T}})$. Equivalently, the t-value may be squared to obtain an F-value proportional to $(\bar{y}_1 - \bar{y}_2)^2 = (\mathbf{c}^{\mathrm{T}}\mathbf{y})^2$. The mean vector $\boldsymbol{\mu}$, which has components $\mu_j = E(y_j)$, has the form $(\mu_1\mathbf{1}_r^{\mathrm{T}}, \mu_2\mathbf{1}_s^{\mathrm{T}})^{\mathrm{T}}$ here and H states that $\mathbf{c}^{\mathrm{T}}\boldsymbol{\mu} = 0$. Under the usual distributional assumption, that the y's are independent normal variates each with variance σ^2, $\mathbf{c}^{\mathrm{T}}\mathbf{y}$ is $N(\mathbf{c}^{\mathrm{T}}\boldsymbol{\mu}, \sigma^2\mathbf{c}^{\mathrm{T}}\mathbf{c})$. (Some normal distribution results are summarized at the end of this section.) Hence, under H, $(\mathbf{c}^{\mathrm{T}}\mathbf{y})^2/\sigma^2\mathbf{c}^{\mathrm{T}}\mathbf{c}$ is χ_1^2 and so will serve as the numerator of F. This outline is deliberately slanted towards generalization: for a general linear hypothesis H: $a_1\mu_1 + a_2\mu_2 = 0$, it applies unchanged but for $\mathbf{c} = (a_1\mathbf{1}_r^{\mathrm{T}}, a_2\mathbf{1}_s^{\mathrm{T}})^{\mathrm{T}}$.

The preceding two-sample case can now be easily extended. Not to be over-ambitious, let us consider the three-sample case, with sample sizes r, s and t, say. Then $\boldsymbol{\mu} = (\mu_1 \mathbf{1}_r^T, \mu_2 \mathbf{1}_s^T, \mu_3 \mathbf{1}_t^T)^T$ and the hypothesis of equal means, H: $\mu_1 = \mu_2 = \mu_3$, can be represented as $\mathbf{c}_1^T \boldsymbol{\mu} = \mathbf{c}_2^T \boldsymbol{\mu} = 0$ with suitable choice of $\mathbf{c}_1$ and $\mathbf{c}_2$. For instance, we might take $\mathbf{c}_1 = (r^{-1} \mathbf{1}_r^T, -s^{-1} \mathbf{1}_s^T, \mathbf{0}_t^T)^T$ (so $\mathbf{c}_1^T \boldsymbol{\mu} = \mu_1 - \mu_2$) and $\mathbf{c}_2 = (r^{-1} \mathbf{1}_r^T, \mathbf{0}_s^T, -t^{-1} \mathbf{1}_t^T)^T$ (so $\mathbf{c}_2^T \boldsymbol{\mu} = \mu_1 - \mu_3$). The separate F-tests corresponding to these two components of H would have numerator chi-squares proportional to $(\mathbf{c}_1^T \mathbf{y})^2$ and $(\mathbf{c}_2^T \mathbf{y})^2$, just as in the two-sample case. To construct a combined F-test we might think of forming a sum, with suitable weights, of $(\mathbf{c}_1^T \mathbf{y})^2$ and $(\mathbf{c}_2^T \mathbf{y})^2$ for the numerator. But here we hit a snag – they are not independent since $\mathbf{c}_1$ and $\mathbf{c}_2$ are not orthogonal; in fact $\mathbf{c}_1^T \mathbf{c}_2 = r^{-2} \mathbf{1}^T \mathbf{1}_r = r^{-1} \neq 0$. Thus a simple addition will not demonstrably yield a chi-square variate for the F-numerator. The solution is to modify $\mathbf{c}_1$ and $\mathbf{c}_2$ to achieve orthogonality, but retaining their interpretation as components of H. For instance, with $\mathbf{c}_1$ as before the choice $\mathbf{c}_2 = (\mathbf{1}_r^T, \mathbf{1}_s^T, -t^{-1}(r+s) \mathbf{1}_t^T)^T$ meets the requirements since then $\mathbf{c}_1^T \mathbf{c}_2 = 0$ and $\mathbf{c}_2^T \boldsymbol{\mu} = r\mu_1 + s\mu_2 - (r+s)\mu_3$; thus $\mathbf{c}_1^T \boldsymbol{\mu} = \mathbf{c}_2^T \boldsymbol{\mu} = 0$ is equivalent to H. Now $U_1 = (\mathbf{c}_1^T \mathbf{y})^2 / \sigma^2 \mathbf{c}_1^T \mathbf{c}_1 \sim \chi_1^2$, $U_2 = (\mathbf{c}_2^T \mathbf{y})^2 / \sigma^2 \mathbf{c}_2^T \mathbf{c}_2 \sim \chi_1^2$, and U_1 and U_2 are independent, hence the numerator for the combined F can be taken as $(U_1 + U_2)/2$. Note that, with normalized vectors $\mathbf{c}_1$ and $\mathbf{c}_2$ (i.e. $\mathbf{c}_1^T \mathbf{c}_1 = \mathbf{c}_2^T \mathbf{c}_2 = 1$), the numerator would have the simple form $\{(\mathbf{c}_1^T \mathbf{y})^2 + (\mathbf{c}_2^T \mathbf{y})^2\}/2\sigma^2$.

We are now in a position to set out the general case. Suppose that the hypothesis to be tested is expressed as H: $\mathbf{c}_j^T \boldsymbol{\mu} = 0$ for $j = 1, \ldots, v$. Assume that (i) any superfluous constraints, implied by the others, have been eliminated (i.e. that the $\mathbf{c}_j$ are linearly independent), and (ii) the $\mathbf{c}_j$ have been **orthonormalized**; by (ii) is meant that the $\mathbf{c}_j$ have been modified, if necessary, to be mutually orthogonal ($\mathbf{c}_j^T \mathbf{c}_k = 0$ for $j \neq k$) and then normalized ($\mathbf{c}_j^T \mathbf{c}_j = 1$ for each j). Then the F-numerator, with v degrees of freedom, is $\sum (\mathbf{c}_j^T \mathbf{y})^2 / v\sigma^2$, with summation over $j = 1, \ldots, v$; $\sum (\mathbf{c}_j^T \mathbf{y})^2$ is the 'sum of squares' corresponding to the hypothesis H. Further, it can be shown that any alternative set of $\mathbf{c}_j$'s which defines the same hypothesis must contain the same number v of vectors and must yield the same sum of squares.

The above discussion has only been concerned with F-numerators. For the denominator we need a χ^2 variate independent of the numerator. It will be evident by now that to construct this we should look for contrasts which (i) are orthogonal to those used in the numerator, and (ii) reflect the required component of error variation only. That it is possible to find such sets, and that this process yields the usual error sums of squares, is part of the standard theory of normal linear models and will not be repeated here.

At last, we can get on to repeated measures. Let $\mathbf{y} = (y_1, \ldots, y_p)^T$ be the vector of observations on an individual and assume a covariance structure as described in section 3.1, i.e. $C(y_j, y_{j'}) = \sigma_\alpha^2 + \delta_{jj'}\sigma^2$. Explicitly, the covariance matrix $\Sigma = V(\mathbf{y})$ of $\mathbf{y}$ has form $\sigma^2 \mathbf{I}_p + \sigma_\alpha^2 \mathbf{J}_p$, where $\mathbf{I}_p$ is a $p \times p$ unit matrix and $\mathbf{J}_p$ is a $p \times p$ matrix of 1's. Thus, Σ has diagonal entries $\sigma_\alpha^2 + \sigma^2$ and off-diagonal entries σ_α^2; the y-components have equal variances $V(y_j) = \sigma_\alpha^2 + \sigma^2$ for each j, and equal correlations, $\rho(y_j, y_{j'}) = \sigma_\alpha^2/(\sigma_\alpha^2 + \sigma^2)$ for every pair $(j \neq j')$. A covariance matrix of such form is said to have **compound symmetry**. Consider now a hypothesis about $\boldsymbol{\mu} = E(\mathbf{y})$ defined by $\mathbf{c}_j^T \boldsymbol{\mu} = 0$ for $j = 1, \ldots, v$. As previously we think of constructing the sum of squares for an F-numerator by adding contributions $(\mathbf{c}_j^T \mathbf{y})^2$, and, as before, we need independent chi-squares. Independence of $\mathbf{c}_j^T \mathbf{y}$ and $\mathbf{c}_k^T \mathbf{y}$ obtains if $0 = C(\mathbf{c}_j^T \mathbf{y}, \mathbf{c}_k^T \mathbf{y}) = \mathbf{c}_j^T \Sigma \mathbf{c}_k^T = \mathbf{c}_j^T(\sigma^2 \mathbf{I}_p + \sigma_\alpha^2 \mathbf{J}_p)\mathbf{c}_k = \sigma^2(\mathbf{c}_j^T \mathbf{c}_k) + \sigma_\alpha^2(\mathbf{c}_j^T \mathbf{1}_p)(\mathbf{c}_k^T \mathbf{1}_p)$. The requirement is certainly met if $\mathbf{c}_j^T \mathbf{c}_k = 0$ (i.e. $\mathbf{c}_j$ and $\mathbf{c}_k$ are orthogonal) and at least one of $\mathbf{c}_j^T \mathbf{1}_p, \mathbf{c}_k^T \mathbf{1}_p$ is zero; a vector $\mathbf{c}$ satisfying $\mathbf{c}^T \mathbf{1}_p = 0$ (i.e. its components having sum zero) is called a **contrast** vector. Hence, if Σ has compound symmetry and the $\mathbf{c}_j$'s are orthogonal contrasts, then the $\mathbf{c}_j^T \mathbf{y}$'s are independent $N(\mathbf{c}_j^T \boldsymbol{\mu}, \sigma^2)$ variates and the F-numerator for testing H may be validly constructed as $\sum(\mathbf{c}_j^T \mathbf{y})^2/\sigma^2$. Actually, there are wider circumstances which yield a valid F-test of which further discussion will appear below.

In analysing repeated measures, particularly those observed on an individual over time, certain linear functions $\mathbf{c}^T \mathbf{y}$ have special interest. First, $\mathbf{c}_0 = p^{-1}\mathbf{1}_p$ yields the mean of the observations $y_1, \ldots, y_p$. Next, the set $\mathbf{c}_1 = (1, -1, 0, \ldots, 0)^T, \mathbf{c}_2 = (1, 0, -1, 0, \ldots, 0)^T, \ldots, \mathbf{c}_{p-1} = (1, 0, \ldots, 0, -1)$ would yield comparisons between the initial response, at time 1, and the succeeding ones. Another set of contrasts commonly employed goes under the name of Helmert: $\mathbf{c}_1 = ((p-1), -1, -1, \ldots, -1)^T, \mathbf{c}_2 = (0, (p-2), -1, \ldots, -1)^T, \ldots, \mathbf{c}_{p-1} = (0, \ldots, 0, 1, -1)^T$. The contrast $\mathbf{c}_j$ here compares μ_j with the average of the following μ's. To examine a possible linear time trend, with measurements equi-spaced in time, take $\mathbf{c}_1 = (1, 2, 3, \ldots, p)^T$; for a quadratic trend take $\mathbf{c}_2 = (1, 4, 9, \ldots, p^2)^T$, and so on. In effect, the observations $y_1, \ldots, y_p$ are replaced, as the primary objects of study, by the set $\mathbf{c}_0^T \mathbf{y}, \ldots, \mathbf{c}_{p-1}^T \mathbf{y}$ of transformed responses. In a computer program the $\mathbf{c}_j$'s are usually orthonormalized in sequence, starting with c_0. Thus c_0 becomes $p^{-1/2}\mathbf{1}_p$, and the c_j's are replaced by equivalent orthogonal contrasts, contrasts since they are each orthogonal to c_0, i.e. $c_j^T\mathbf{1}_p = 0$ for $j \geqslant 1$.

A 'complete set of orthonormal contrasts' means a set $(c_0,\ldots,c_{p-1})$ like those described above. Strictly speaking $c_0 = p^{-1}\mathbf{1}_p$ is not a contrast as such, but is usually included in the set being, by definition, orthogonal to any proper contrast. A complete set will then contain c_0 (as $p^{-1}\mathbf{1}_p$, or $p^{-1/2}\mathbf{1}_p$ in its normalized version) plus $p-1$ proper contrasts each orthogonal to c_0 and to each other. All p dimensions are then accounted for.

Since compound symmetry of Σ has been a vital ingredient of this section, we mention here a test for it (Rouanet and Lepine, 1970). Let $\mathbf{S}$ be the sample covariance matrix, e.g. for a single group of n individuals $\mathbf{S}$ has (j, k)th element $S_{jk} = (n-1)^{-1}\sum_i(y_{ij} - \bar{y}_j)(y_{ik} - \bar{y}_k)$ where $\bar{y}_j = n^{-1}\sum y_{ij}$ and summations are over $i = 1,\ldots,n$. Let $a = p^{-1}\sum_j S_{jj}$ and $b = \sum_{j \neq k} S_{jk}/p(p-1)$ be the averages of the diagonal and off-diagonal entries of $\mathbf{S}$, and let $s_1 = a + (p-1)(a-b)bp - 1$. Then $(k-n)|s_1|^{-1}\log(\det\mathbf{S})$ is approximately χ_v^2, with $v = p(p+1)/2 - 2$, for large n under the hypothesis of compound symmetry.

3.5.1 Distributional results

We note here some basic results used in this section. Let $\mathbf{y}(p \times 1)$ have mean $\boldsymbol{\mu}(p \times 1)$ and covariance matrix $\Sigma(p \times p)$, i.e. $E(y_j) = \mu_j$ and $C(y_j, y_k) = \Sigma_{jk}$. Then $E(\mathbf{c}^T\mathbf{y}) = \mathbf{c}^T\boldsymbol{\mu}$ and $V(\mathbf{c}^T\mathbf{y}) = \mathbf{c}^T\Sigma\mathbf{c}$, where $\mathbf{c}$ is $p \times 1$, and $C(\mathbf{c}_1^T\mathbf{y}, \mathbf{c}_2^T\mathbf{y}) = \mathbf{c}_1^T\Sigma\mathbf{c}_2$.

Now suppose that $\mathbf{y}$ is $N_p(\boldsymbol{\mu}, \Sigma)$, i.e. that $\mathbf{y}$ has a p-variate normal distribution with mean $\boldsymbol{\mu}$ and covariance matrix Σ. If $\Sigma_{jk} = 0$ then y_j and y_k are uncorrelated; in the case of the normal distribution this means that they are independent. Thus, if Σ is diagonal the y_j's are all independent. The random variable $\mathbf{c}^T\mathbf{y}$ is $N(\mathbf{c}^T\boldsymbol{\mu}, \mathbf{c}^T\Sigma\mathbf{c})$. If $\Sigma = \sigma^2\mathbf{I}_p$, so that the y_j's are independent each with variance σ^2, and $\mathbf{c}_1^T\mathbf{c}_2 = 0$, so that $\mathbf{c}_1$ and $\mathbf{c}_2$ are orthogonal, then $\mathbf{c}_1^T\mathbf{y}$ and $\mathbf{c}_2^T\mathbf{y}$ are independent since their covariance is $\mathbf{c}_1^T\Sigma\mathbf{c}_2 = \sigma^2\mathbf{c}_1^T\mathbf{c}_2 = 0$.

Let Z be $N(\mu_Z, \sigma_Z^2)$, i.e. univariate normal, then $(Z - \mu_Z)/\sigma_Z$ is $N(0, 1)$, standard normal, and $(Z - \mu_Z)^2/\sigma_Z^2$ is χ_1^2. Let U_1 and U_2 be independent with distributions χ_a^2 and χ_b^2. Then (i) $U_1 + U_2$ is χ_{a+b}^2, and (ii) $a^{-1}U_1/b^{-1}U_2$ is $F_{a,b}$. If $a = 1$ and $Z = U_1^{1/2}$ in (ii) we have Z as $N(0, 1)$ independent of U_2 and $Z/(b^{-1}U_2)^{1/2}$ as $\sqrt{F_{1,b}}$, i.e. the t_b-distribution.

3.6 Relaxing assumptions: univariate, modified univariate, or multivariate tests?

Compound symmetry is a special case of a more general situation under which the simple F-tests are valid. In this section we examine this more general situation and also look at modifications of the straightforward F-statistic which permit it to be applied even when the necessary conditions are not satisfied.

Let us first take the simple case of a single group of subjects, each measured on p occasions. We can apply $(p-1)$ orthogonal contrasts to these p occasions, such that the set of transformed responses summarizes the pattern of differences over the occasions. For convenience we shall assume that the orthogonal contrasts are also normalized – they are orthonormal contrasts. Then the mean-square ratios derived by the univariate approach follow exact F-distributions if and only if the covariance matrix of the orthonormal contrasts has equal variances and zero covariances, that is, if and only if the covariance matrix is a scalar multiple of the identity matrix. This condition is called the **sphericity** or **circularity** condition. It can be expressed in a number of alternative ways. For example, one way which is sometimes convenient is that all the variances of pairwise differences between variables, $V(x_i - x_j)$, are equal. When put like this, one can easily see where the compound symmetry assumption comes from as follows. We need $V(x_i - x_j) = $ constant for all i and j. Now

$$V(x_i - x_j) = V(x_i) + V(x_j) - 2C(x_i, x_j) \tag{3.6.1}$$

Thus, it is clearly the case that $V(x_i - x_j) = $ constant if $V(x_i) = $ constant and $C(x_i, x_j) = $ constant for all i and j, i.e. if compound symmetry holds. However, we can now also see that there will be other conditions under which (3.6.1) is true without the individual components being constant. Thus, sphericity is a more general condition than compound symmetry.

Each term on the diagonal of the covariance matrix of contrasts estimates the denominator mean square in an F-test of that contrast. Thus, when sphericity holds, they each estimate the same thing. In such a case one can get a better estimate of the common value by averaging them. This is exactly what the usual univariate test does, and hence it is sometimes called the 'averaged F-test'. If sphericity does not hold one can see that one will be averaging different things.

The resulting tests on individual contrasts may then have denominators which are too large or too small, thus biasing the test. We briefly return to individual contrasts below. Note that if one is only interested in some subset of derived variables one can either restrict the sphericity requirement to those variables or one can require all $p - 1$ variables to satisfy sphericity. The test of the latter has more degrees of freedom in the denominator sum of squares and so will generally be more powerful, but it is also more likely to deviate from sphericity. We illustrate how this sort of situation might arise below.

Turning now to more general cases, let us first extend the design to several groups, the subjects within each group each being measured on several occasions. Then, for the univariate F-tests to be valid, in addition to the sphericity condition we must also assume that the covariance matrices of the orthonormal contrasts are identical across the several groups.

The next extension is to the situation where there is some structure on the occasions at which the measurements occur. For example, in Example 3.2 each subject was measured on two eyes under each of four conditions (lenses). There was thus a two-by-four structure on the repeated measures. This is an example of the situation described above. One can make a global sphericity assumption, with the advantage of many degrees of freedom in the F-denominator. In this case, however, the orthonormal contrasts can be usefully partitioned into three groups: one group refers to the main effect *eye*, one to the main effect *lens*, and the third group to *eye* by *lens* interaction. While an overall test of differences between these eight measurements would require a global sphericity assumption, such a test would rarely be of interest. It is much more likely that one would be interested in the effect of eyes, of lenses, and of the interaction between the two, as separate issues as discussed earlier. In this case, for each effect it is only necessary to assume sphericity for the contrasts concerned. Thus the test on lenses only assumes that the covariance matrix for the orthonormal contrasts describing the lens effect is spherical, and so on. Clearly this is a much less restrictive assumption than global sphericity. It is, of course, still necessary to assume equality of the covariance matrices across groups if more than one group is involved.

In general, then, for the univariate tests to be valid, we must assume equality of the covariance matrices of the orthonormal contrasts across groups, and also that this common covariance matrix is spherical. Given this situation, naturally investigators have considered ways in which these necessary conditions might be tested. First let us look at tests that the groups have a common covariance matrix.

Box (1949) has given the following generalization of the Bartlett test for equality of G variances. Let $\mathbf{S}_g$ be the sample covariance matrix for the gth group of n_g individuals, and let $\mathbf{S}$ be the pooled covariance matrix:

$$\mathbf{S} = \sum_g (n_g - 1)\mathbf{S}_g / \sum_g (n_g - 1)$$

The test statistic is then αM, where

$$M = \sum_g (n_g - 1) \log (\det \mathbf{S} / \det \mathbf{S}_g)$$
$$\alpha = 1 - \tfrac{1}{6}(2p^2 + 3p - 1)(p + 1)^{-1}(G - 1)^{-1}$$
$$\cdot \{ \sum_g (n_g - 1)^{-1} - [\sum_g (n_g - 1)]^{-1} \}$$

αM is approximately distributed as a chi-square with $(G - 1)p(p + 1)/2$ degrees of freedom under the null hypothesis. This approximation seems to be good for G and p less than 5 with each n_g greater than 20. Box (1949) also describes an F approximation which seems to be satisfactory when none of the n_g are less than 10. Let

$$f_1 = \tfrac{1}{2}p(p + 1)(G - 1), \qquad f_2 = (f_1 + 2)/\{ A - (1 - \alpha)^2 \},$$
$$b = f_1 / (\alpha - f_1 / f_2)$$
$$A = \tfrac{1}{6}(p - 1)(p + 2)(G - 1)\{ \sum_g (n_g - 1)^{-2} - [\sum_g (n_g - 1)]^{-2} \}$$

where α and M are as above. The test statistic is then M/b, referred to F_{f_1, f_2}.

Once one is satisfied that the covariance matrices are identical across groups one can proceed to test that they are spherical. The more popular test for this, due to Mauchly (1940), has test statistic $W = (\det \mathbf{R}) / |q^{-1} \operatorname{trace} \mathbf{R}|^q$, where $q = p - 1$, $\mathbf{R} = \mathbf{C}\mathbf{S}\mathbf{C}^{\mathrm{T}}$ and $\mathbf{C}$ is a $q \times p$ orthonormal contrast matrix. Tables for W are given in Nagarsenker and Pillai (1972) and in Kres (1983); alternatively, $\{ \Sigma(n_g - 1) - v \} \log W$ is asymptotically distributed as χ_v^2 with $v = \tfrac{1}{2}(q - 1)(q + 2)$.

Tests of variances and covariances are known to be sensitive to non-normality and Huynh and Mandeville (1979) discuss robustness of the Mauchly test to such departure. From simulation studies they conclude that: (a) the Mauchly W criterion tends to err on the conservative side for light-tailed distributions, the discrepancy between the empirical type I error and the nominal significance level being greater for large samples and for small α; (b) with heavy-tailed distributions the reverse is true, with many more sample values of W exceeding the critical value for the nominal significance level. Again, the situation becomes worse with increasing n and decreasing α. Sometimes the F_{max} criterion, the ratio of the largest to the smallest variance, is also used as a test for sphericity. The result, however, will depend on the particular orthonormal transformation chosen.

Of course, one should be wary of placing too much emphasis on tests of assumptions – one does not want to reject a robust analysis which would have been relatively accurate on the basis of an overly sensitive test of assumptions. Perhaps this is partly where the notion of statistics as an 'art' comes in! We briefly discuss the overall strategy on the basis of a sphericity test below, as well as the question of the power of the Mauchly test.

So far we have explored conditions under which the simple univariate F-tests are valid and have described tests for examining those conditions. There are two other obvious questions which must now be answered. One is: what is the effect of using the ordinary F-tests when sphericity does not hold? The other is: what do we do if the assumptions seems unjustifiable? We discuss these in turn.

If sphericity does not hold then the actual level of significance of the univariate F-tests will exceed the nominal level – too many true null hypotheses will be rejected. For tests on complete sets of orthonormal contrasts this will become clearer when we discuss the ε correction factor below. Rouanet and Lepine (1970), Mitzel and Games (1981), and Boik (1981) all discuss the effect of non-sphericity on single contrasts. Boik gives numerical examples and concludes: 'On the whole, the ordinary F-tests have nothing to recommend them. The type I error rate is out of control and the power to detect an effect is not a monotonic function of the reliability of the effect.' As a consequence Boik recommends the use of error terms appropriate for the constrast(s) in question and not obtained by averaging over all $(p - 1)$ constrasts. Rouanet and Lepine (1970)) recommend a more flexible approach involving examining all relevant statistics.

Now we turn to the second question raised above: what should one do if the sphericity assumption does not hold? The two most common approaches are either to adopt a multivariate approach (discussed in Chapter 4) or to adapt the univariate statistics. Since the multivariate approach makes no assumptions at all about the form of the (common) covariance matrix, it is always a legitimate procedure to adopt. Its disadvantage lies in a relatively low power when the sphericity assumptions are satisfied compared to the univariate approach. Rouanet and Lepine (1970) state that, when the sphericity conditions are true, the multivariate test statistic underestimates the theoretical ratio of mean squares which is estimated by the univariate approach. Moreover, though the two tests have the same numerator degrees of freedom, the univariate test has greater denominator degrees of freedom. Thus, ignoring the possibility of sphericity and always adopting the multivariate approach is not the best general policy.

Adaptations of the univariate method to the non-spherical situation hinge around ε, a measure of deviation from sphericity. If Σ_c is the covariance matrix of the $(p - 1)$ orthonormal contrasts, then

$$\varepsilon = (\text{trace}\,\Sigma_c)^2/(p - 1)\,\text{trace}\,(\Sigma_c^2) = (\textstyle\sum_j \theta_j)^2/(p - 1)\textstyle\sum_j \theta_j^2$$

where the θ_j are the $(p - 1)$ eigenvalues of Σ_c. The overall F-test for an occasions effect in the case of G groups with n individuals and p measures involves the $F_{p-1,(p-1)(n-G)}$ distribution. When nonsphericity is present the distribution of the variance ratio is approximated by the $F_{\varepsilon(p-1),\varepsilon(p-1)(n-G)}$ distribution, i.e. numerator and denominator degrees of freedom are each multiplied by ε. It is easy to see from the above definition of ε that when sphericity holds $\varepsilon = 1\cdot0$, since then the $(p - 1)$ eigenvalues are equal, and that otherwise $\varepsilon < 1.0$. Thus, when sphericity holds the modified test becomes the standard one. When $\varepsilon < 1\cdot0$ the reduction in degrees of freedom is, of course, likely to yield nonintegral values so that interpolation will be needed in the F-tables. This only refers to tests of occasions, tests comparing groups remain unaltered.

Since ε will not be known it will have to be estimated from the data, and this introduces a complication – how will the inaccuracy in the estimate of ε influence the results? Greenhouse and Geisser (1959) avoid this problem by suggesting the conservative approach of replacing ε by its smallest possible value, namely $1/(p - 1)$. In Example 3.1, for instance, this changes the degrees of freedom in the F-test for the period effect from (5, 60) to (1, 12). The strategy outlined by Greenhouse and Geisser is as follows. First conduct the F-test with unreduced degrees of freedom. If the F-value is smaller than the critical value, i.e. not significant, then stop. If the result is significant then conduct the conservative test, reducing the degrees of freedom by a factor $1/(p-1)$. If this is significant then stop – the null hypothesis is rejected. If this is not significant then estimate ε from the observed covariance matrix and conduct the approximate test.

Collier *et al.* (1967), Stoloff (1970), Mendoza, Toothaker and Nicewander (1974), Wilson (1975), and Huynh (1978) have explored the effect on type I error of adjusting the degrees of freedom by $\hat{\varepsilon} = \varepsilon(\mathbf{S})$ computed from the sample covariance matrix $\mathbf{S}$. It seems that, for ε greater than 0.75 and n less than $2p$, $\hat{\varepsilon}$ may be seriously biased so that it tends to overcorrect the degrees of freedom and produce a conservative test. Huynh and Feldt (1976) have therefore derived an alternative estimator of ε which is less biased and less dependent on large n when ε is greater than about 0.75. Their estimator, based on using the ratio of unbiased estimators of the numerator and denominator of ε, is $\tilde{\varepsilon} = \min(1, a/b)$ where $a = n(p-1)\hat{\varepsilon} - 2$ and $b = (p-1)\{n - G - (p-1)\hat{\varepsilon}\}$.

If the covariance matrix of the orthonormal contrasts satisfies sphericity one would naturally conduct the unmodified F-test. When there is non-sphericity, or when one has no sound theoretical reason for supposing that sphericity holds, however, one is confronted by the question of whether or not to perform a sphericity test and choose the sequel according to the result. We briefly referred to this issue above. To answer this question Rogan, Keselman and Mendoza (1979) carried out a Monte Carlo study, comparing the power and type I error of a Mauchly test followed by either (i) an unadjusted univariate F-test, or (ii) an $\hat{\varepsilon}$- (or $\tilde{\varepsilon}$-) adjusted F-test, or a multivariate test. They found that the result differed little from consistently using an $\hat{\varepsilon}$- (or $\tilde{\varepsilon}$) adjusted F-test, or a multivariate test without first testing for sphericity. They found no clear-cut rule for choosing between these three latter options – there were differences in power and robustness according to the value of ε, but of course a researcher will not know ε.

We remarked above that Rouanet and Lepine (1970) pointed out that the multivariate test is likely to be less powerful than the unmodified F-test when sphericity holds. However, these authors also suggest that the multivariate approach is often not much less powerful than the univariate one not pooling the denominator sum of squares. And, of course, for a test on a single derived variable Hotelling's T^2 test and the univariate test with unpooled sums of squares are identical. Boick (1981) also reports that 'a very small departure from sphericity is sufficient to produce a very large bias in the (unmodified) F-tests', and that Mauchly's criterion provides little information for detecting minute departures from sphericity. For instance, with $\alpha = 0.05$, $p = 4$, $G = 3$, $n = 39$ and $\varepsilon = 0.96$, Keselman et al. (1980) show the power of Mauchly's test to be 0.15.

The final conclusion to the burning question which will be uppermost in everyone's mind (namely: what on earth do I do?) is thus, if you have strong prior reasons for believing that sphericity holds then perform the standard unmodified F-test, and if not then perform either a modified F-test (with $\hat{\varepsilon}$ or $\tilde{\varepsilon}$) or a multivariate test. There seems to be little to choose between the modified F-tests and the multivariate test, except when $p > n$. In the latter case the covariance matrix is singular so that the matrix inversion necessary for the multivariate test cannot be performed.

Example 3.4

This data set and its analysis is described in detail in Hand and Taylor (1987, p. 117). Here we merely consider the issues discussed in this section.

The data consist of two groups of subjects, those with severe and those with moderate dependence on alcohol. Each subject had their salsolinol (an alkaloid with a similar structure to heroin) measured on four days. Here we consider the question of whether the two groups showed different salsolinol excretion patterns over time. The data are given in Table 3.7. For reasons explained in Hand and Taylor (1987), we work with log-transformed data.

Both the test of overall profile shape and the test of shape-by-group interaction (i.e. of whether the groups evolve differently over time) involve the sphericity assumption. This produced a p-value of 0.45 (using BMDP2V), so that the hypothesis was not rejected. Although the Greenhouse–Geisser estimate $\hat{\varepsilon}$ was only 0.77, it merely altered the p-values for the F-tests from 0.48 to 0.45 for the shape test and from 0.63 to 0.59 for the shape-by-group test. The Huynh–Feldt estimate $\tilde{\varepsilon}$ took its maximum value of 1.0, so that the degrees of freedom for the F-test remained unaltered.

Table 3.7. *Salsolinol excretion data (in mmol)*

Group	Day 1	Day 2	Day 3	Day 4
1	0·33	0·70	2·33	3·20
1	5·30	0·90	1·80	0·70
1	2·50	2·10	1·12	1·01
1	0·98	0·32	3·91	0·66
1	0·39	0·69	0·73	2·45
1	0·31	6·34	0·63	3·86
2	0·64	0·70	1·00	1·40
2	0·73	1·85	3·60	2·60
2	0·70	4·20	7·30	5·40
2	0·40	1·60	1·40	7·10
2	2·60	1·30	0·70	0·70
2	7·80	1·20	2·60	1·80
2	1·90	1·30	4·40	2·80
2	0·50	0·40	1·10	8·10

The multivariate test of shape gave $p = 0.32$ and of shape-by-group interaction, $p = 0.90$. □

Example 3.2 (contd)

The orthonormal contrasts in this example produce three sets of variables. One set describes the eye effect, one the lens strength effect, and one the interaction of these two. Thus, as described above, there are three sets of variables to be considered with respect to the sphericity assumption. (Since there is but a single group we need not concern ourselves with testing the homogeneity of the covariance matrices across groups.) However, this example illustrates an important special case. There are only two eyes, so that there is only a single orthonormal contrast for this effect. The 'covariance matrix' is thus a scalar, satisfying sphericity trivially. This means that it is unnecessary to conduct a sphericity test for the eye effect. The situation is the same as in the paired comparison t-test; in fact, it is another way of looking at the test. This leaves us with two sets of variables, one for lens strength and one for the eye-by-lens strength interaction. The results of the various tests are illustrated below.

Lens strength. The sphericity test gave $p = 0.0046$ (in PMDP2V) and the F-statistic had value 2·25. The Greenhouse–Geisser estimate

$\hat\varepsilon$ was 0·50 and the Huynh–Feldt estimate $\tilde\varepsilon$ was 0·62. The following table lists the various F-tests.

Test	d.f. used	p-value
Unadjusted F	3, 18	0·118
Conservative F	1, 6	0·184
$\hat\varepsilon$-adjusted F	$3\hat\varepsilon$, $18\hat\varepsilon$	0·167
$\tilde\varepsilon$-adjusted F	$3\tilde\varepsilon$, $18\tilde\varepsilon$	0·153

The multivariate test in this case yields $p = 0·006$; all the multivariate criteria (section 4.4) yield the same result because there is only one group.

Eye by lens strength. The sphericity test gave $p = 0.23$ (in BMDP2V) and the F-statistic had value 1·06. The Greenhouse–Geisser estimate $\hat\varepsilon$ was 0·55 and the Huynh–Feldt $\tilde\varepsilon$ was 0·73. The following table gives the F-tests in this case.

Test	d.f. used	p-value
Unadjusted F	3, 18.	0·393
Conservative F	1, 6	0·343
$\hat\varepsilon$-adjusted F	$3\hat\varepsilon$, $18\hat\varepsilon$	0·370
$\tilde\varepsilon$-adjusted F	$3\tilde\varepsilon$, $18\tilde\varepsilon$	0·382

The multivariate test yields $p = 0·673$; all the multivariate criteria yield the same result because there is only one group. $\square$

3.7 Further reading

The general topic of analysis of variance is well covered, if not always covered well, in the literature. Standard works include Anderson and Bancroft (1952), Cochran and Cox (1957), Scheffé (1959), Winer (1971) and Searle (1971). These books treat random effects in Anova to various levels of sophistication. Source works for fixed, mixed and random models, with which section 3.1 here is concerned, include Eisenhart (1947), Wilk and Kempthorne (1955), and Scheffé (1956a, b). A recent discussion of these models is given in Dawid (1988) (the author coined the unforgettable title 'The Great Mixed Model Muddle' for his previous talks on the subject).

The calculation of expected mean-squares, touched on in section 3.3, is tackled by Scheffé (1959, Chapter 8), Winer (1971, Chapters 4 and 7) and Searle (1971, Chapters 9 and 10), among others; the last named describes methods more sophisticated than those adopted here.

The treatment of contrasts in section 3.5 owes much to Scheffé (1959, Chapters 1 and 2). Distributional results are given a detailed and extensive treatment by Searle (1971, Chapter 2).

Some of the more important articles on the material of section 3.6 have been cited in the text. A great deal of the literature has been published in psychological journals, in particular *Psychological Bulletin* and *Psychometrika*. The topic is one that has stirred up much interest and disagreement amongst behavioural scientists. Rouanet and Lepine (1970) give a very lucid description of the issues.

John (1972) and Sugiura (1972) have developed an alternative to Mauchly's W-test of sphericity, the former showing it to be the locally best invariant test. Grieve (1984) shows that W is a function of ε and can be written as $U = \text{trace } \mathbf{S}^2/(\text{trace } \mathbf{S})^2 = 1/q\varepsilon$ where q, the number of contrasts under test, is less than p. No uniformly most powerful test exists but U provides the locally most powerful one. Grieve (1984) also gives tables directly for ε. Harris (1984) obtains a multisample sphericity test using the efficient scores criterion: $V = \frac{1}{2}(nq)^2 (\sum_g n_g \text{ trace } \mathbf{S}_g^2)/(\sum_g n_g \text{ trace } \mathbf{S}_g)^2$ is approximately χ_v^2 under sphericity, with $v = \frac{1}{2}q(q+1)G - 1$, $q < p$.

Huynh (1978) extends the work of Huynh and Feldt (1976) on improved ε-estimation to the case of unequal covariance matrices over groups, but concludes that the extension is computationally rather complex and that, in many situations, $\tilde{\varepsilon}$ performs just as well. Thomas (1983) has extended the sphericity conditions to the case of multiple measurements made on each occasion. He calls this 'multivariate mixed model' analysis.

CHAPTER 4

Multivariate analysis

4.1 Models without special covariance structure

From now on there is a shift of viewpoint in that the p measurements $(y_{i1}, \ldots, y_{ip})$ on an individual will be treated as a single vector multivariate observation $\mathbf{y}_i$ rather than as a set of separate univariate observations. This sounds like just shifting the furniture around, but it opens up the route to the more advanced treatment of the subject. The primary interest is still in inference about the means μ_{ij} of (3.1.1) as in Anova, but these parameters are now treated as components of the $p \times 1$ vector $\boldsymbol{\mu}_i = E(\mathbf{y}_i)$. The mean-squares of Anova, which represent components of univariate variation, are now replaced by covariance matrices which represent components of multivariate covariation. The general model on which the analysis is based is $\mathbf{y}_i = \boldsymbol{\mu}_i + \mathbf{e}_i$ for individual i, corresponding to (3.1.1), where the errors $\mathbf{e}_i$ are independent with means 0 and covariance matrices $V(\mathbf{e}_i) = \Sigma$; thus Σ is $p \times p$ with (j, k)th element $C(e_{ij}, e_{ik}) = \Sigma_{jk}$. Note the implication here that $V(\mathbf{y}_i) = V(\mathbf{e}_i) = \Sigma$ is the same for all i.

In the context of repeated measurements the phrase 'multivariate methods' is often used to mean that no particular structure is assumed for Σ. This is in contrast to (3.1.2) where Σ_{jk} is assigned a special form. However, in subsequent chapters, where Σ will be structured, the analysis is no less multivariate in the wider statistical sense. While it is robust not to assume knowledge of the covariance structure, this can result in rather weak inference in the sense that too many degrees of freedom are used up in estimating the covariance parameters, leaving too few for the parameters of interest. For instance, even with only $p = 5$ repeated measurements there are 15 distinct covariance parameters in Σ.

4.2 Hotelling's T^2

The familiar t-test for the single sample case is constructed as follows. Suppose that there are univariate observations $y_1, \ldots, y_n$ forming a sample of size n from $N(\mu, \sigma^2)$, a normal distribution with mean μ and variance σ^2. The sample mean $\bar{y}$ and sample variance s^2 are independently distributed, $\bar{y}$ as $N(\mu, \sigma^2/n)$ and $(n-1)s^2/\sigma^2$ as χ^2_{n-1}. To test the null hypothesis $H_0: \mu = \mu_0$ the statistic $t = n^{1/2}(\bar{y} - \mu_0)/s$ is used.

Suppose now that the data are multivariate, comprising vector observations $\mathbf{y}_1, \ldots, \mathbf{y}_n$ from $N_p(\boldsymbol{\mu}, \boldsymbol{\Sigma})$, a p-dimensional normal distribution with mean $\boldsymbol{\mu}$ and covariance matrix $\boldsymbol{\Sigma}$. The sample mean vector $\bar{\mathbf{y}}$ and sample covariance matrix $\mathbf{S}$, where $S_{jk} = (n-1)^{-1}\sum_{i=1}^{n}(y_{ij} - \bar{y}_j)(y_{ik} - \bar{y}_k)$, are independently distributed, $\bar{\mathbf{y}}$ as $N_p(\boldsymbol{\mu}, \boldsymbol{\Sigma}/n)$ and $(n-1)\mathbf{S}$ as $W_p(n-1, \boldsymbol{\Sigma})$ (the p-dimensional Wishart distribution with $n-1$ degrees of freedom). The multivariate analogue of the t-statistic, whose square is $t^2 = n(\bar{y} - \mu_0)^2/s^2$, is $T^2 = n(\bar{\mathbf{y}} - \boldsymbol{\mu}_0)^{\mathrm{T}}\mathbf{S}^{-1}(\bar{\mathbf{y}} - \boldsymbol{\mu}_0)$, known as Hotelling's T^2.

Consider now the two-sample univariate case, involving independent observations from $N(\mu_1, \sigma^2)$ and $N(\mu_2, \sigma^2)$. The t-test for equality of the means μ_1 and μ_2 is based on $t = (n_1^{-1} + n_2^{-1})^{-1/2}(\bar{y}_1 - \bar{y}_2)/s$ where $(n_1 + n_2 - 2)s^2 = (n_1 - 1)s_1^2 + (n_2 - 1)s_2^2$; n_1 and n_2 are the sample sizes, $\bar{y}_1$ and $\bar{y}_2$ are the sample means, s_1^2 and s_2^2 are the sample variances, and s^2 is the pooled sample variance. Hotelling's T^2 for the corresponding multivariate situation is defined analogously as $T^2 = (n_1^{-1} + n_2^{-1})^{-1}(\bar{\mathbf{y}}_1 - \bar{\mathbf{y}}_2)^{\mathrm{T}}\mathbf{S}^{-1}(\bar{\mathbf{y}}_1 - \bar{\mathbf{y}}_2)$ where $(n_1 + n_2 - 2)\mathbf{S} = (n_1 - 1)\mathbf{S}_1 + (n_2 - 1)\mathbf{S}_2$.

The general definition of a Hotelling's T^2 statistic is $T^2 = v\mathbf{z}^{\mathrm{T}}\mathbf{A}^{-1}\mathbf{z}$ where $\mathbf{z}$ and $\mathbf{A}$ are independently distributed as $N_q(\mathbf{0}, \boldsymbol{\Gamma})$ and $W_q(v, \boldsymbol{\Gamma})$, respectively. It can be shown that $(v - q + 1)T^2/vq$ has variance ratio F-distribution $F_{q, v-q+1}$ with the indicated degrees of freedom.

Example 4.1

Recall the shabby treatment meted out to the data in Example 2.1, where Groups 1 and 2 were compared in terms of a mere t-test between the means at time 5 minutes. We can improve on this a bit now with a T^2-test between these groups for several times simultaneously. Let us take times 1, 5 and 10 minutes for illustration. The sample sizes are $n_1 = 8$ and $n_2 = 6$, the sample mean vectors are $\bar{\mathbf{y}}_1 = (6{\cdot}83, \ 6{\cdot}81, \ 6{\cdot}39)$ and $\bar{\mathbf{y}}_2 = (7{\cdot}73, \ 6{\cdot}98, \ 8{\cdot}35)$, and the sample covariance matrices are

$$
\mathbf{S}_1 = \begin{pmatrix} 14{\cdot}1 & \cdot & \cdot \\ 12{\cdot}7 & 12{\cdot}4 & \cdot \\ 12{\cdot}5 & 11{\cdot}9 & 11{\cdot}7 \end{pmatrix} \quad \text{and} \quad \mathbf{S}_2 = \begin{pmatrix} 10{\cdot}8 & \cdot & \cdot \\ 10{\cdot}4 & 12{\cdot}7 & \cdot \\ 12{\cdot}5 & 12{\cdot}5 & 16{\cdot}2 \end{pmatrix}.
$$

The pooled sample covariance matrix is computed as $S = (7S_1 + 5S_2)/12$ and then $T^2 = 6.35$ using the formula given above. Hence $F_{3,10} = 1.76$ ($p > 0.10$), noting that $q = 3$ and $v = n_1 + n_2 - 2 = 12$ here.

Suppose that we were to perform the same test, between Groups 1 and 2, but based on the observations at all the times after 0 minutes. In this case $v = 12$ as before, but now $p = 10$ and we would obtain an $F_{10,3}$ statistic. As p increases the degrees of freedom, $v - p + 1$, for the denominator of F decrease and reach 0 when $p = n_1 + n_2 - 1$, at which point S becomes singular. This illustrates the point made in section 4.1 that freedom from covariance assumptions has to be paid for, in this case by loss of test power.

Let us go back to times 1, 5 and 10 minutes. The test above, yielding $T^2 = 6.35$, is of the hypothesis that the mean vectors μ_1 and μ_2 for the two groups are equal, i.e. that the μ-profiles are the same. A slightly weaker hypothesis, which is often more realistic in many applications, is that the μ-profiles are parallel, i.e. that the μ_1 and μ_2 curves differ only by a constant vertical shift. This may be expressed as $H\mu_1 = H\mu_2$ where $H = \begin{pmatrix} 1 & -1 & 0 \\ 0 & 1 & -1 \end{pmatrix}$: thus the action of H on the μ's is to take differences between successive components. Under this hypothesis $H(\bar{y}_1 - \bar{y}_2)$ has mean 0, and its covariance matrix is $H\{V(y_1) + V(y_2)\}H^T = (n_1^{-1} + n_2^{-1})H\Sigma H^T$. In the general definition of T^2 given above take $z = (n_1^{-1} + n_2^{-1})^{-1/2}H(\bar{y}_1 - \bar{y}_2)$ and $A = H[(n_1 - 1)S_1 + (n_2 - 1)S_2]H^T$. Then z and A satisfy the conditions with $q = 2, v = n_1 + n_2 - 2 = 12$ and $\Gamma = H\Sigma H^T$. Thus $T^2 = vz^TA^{-1}z$ and the numerical value comes out as $T^2 = 6.31$ for these data; correspondingly, $F_{2,11} = 2.89$ ($0.05 < p < 0.10$). It can be shown that the T^2 value will be the same under equivalent, alternative versions of H. For instance, the choice $H = \begin{pmatrix} -\frac{1}{2} & 0 & \frac{1}{2} \\ 1 & -2 & 1 \end{pmatrix}$ also leads to $T^2 = 6.31$.   □

Example 4.2

The twelve patients in Example 3.3 were observed during three phases (two occasions pre-treatment, three occasions during, and two occasions post-treatment). It was assumed in the Anova of section 3.2 that there were no changes of mean level within phases. This explains the appearance of just σ^2 in Table 3.6 as the E(m.s.) for *occasions-within-phases*. Actually, we can test this assumption using the results in that table by $F_{4,44} = (0.2034/4) \div (2.0522/44) = 1.09$; had this F been large the model would have been thrown into doubt.

To perform a T^2 test of the same assumption we express it in terms of the components of μ, the mean vector of dimension $p = 7$. The hypothesis is then that $\mu_1 = \mu_2$, $\mu_3 = \mu_4 = \mu_5$, and $\mu_6 = \mu_7$. This may be expressed succinctly as $H\mu = 0$ where H is the 4×7 matrix

$$H = \begin{pmatrix} 1 & -1 & 0 & 0 & 0 & 0 & 0 \\ 0 & 0 & 1 & -1 & 0 & 0 & 0 \\ 0 & 0 & 1 & 0 & -1 & 0 & 0 \\ 0 & 0 & 0 & 0 & 0 & 1 & -1 \end{pmatrix}.$$

In the general definition of T^2 take $z = n^{1/2}H\bar{y}$ and $A = (n-1)HSH^T$ where $\bar{y}$ and S are the sample mean vector and sample covariance matrix. Then z and A satisfy the conditions with $n = 12$, $q = 4$, $v = n - 1$ and $\Gamma = H\Sigma H^T$, Σ being the underlying covariance matrix of y. Thus $T^2 = n\bar{y}^T H^T (HSH^T)^{-1} H\bar{y} = 14.8$ and $F_{4,8} = 2.69$ ($p > 0.10$). Notice how many fewer degrees of freedom we now have in the denominator compared with the Anova test, 8 compared with 44. In general terms, the comparison is of $n - (p - P)$ with $(n-1)(p - P)$, where P is the number of phases; the numerator degrees of freedom are p $-$ P in both tests. $\square$

Example 4.3

In the pilot trial of Example 2.2 there are $n = 5$ subjects and $p = 8$ occasions partitioned into two similar phases, one with each of drugs A and B. Correspondingly, let $\mu = (\mu_A, \mu_B)$. The null hypothesis of no difference between drugs is $\mu_A = \mu_B$, and, in general linear model terms, this is $H\mu = 0$ where

$$H = \begin{pmatrix} 1 & 0 & 0 & 0 & -1 & 0 & 0 & 0 \\ 0 & 1 & 0 & 0 & 0 & -1 & 0 & 0 \\ 0 & 0 & 1 & 0 & 0 & 0 & -1 & 0 \\ 0 & 0 & 0 & 1 & 0 & 0 & 0 & -1 \end{pmatrix}.$$

More briefly, $H = (I_4, -I_4)$, where I is the 4×4 unit matrix, so $H\mu = \mu_A - \mu_B$. To construct the appropriate T^2 take $z = n^{1/2}Hy = n^{1/2}(\bar{y}_A - \bar{y}_B)$, and $A = (n-1)HSH^T = (n-1)(S_{AA} - 2S_{AB} + S_{BB})$. Here the sample mean vector $y = (\bar{y}_A, \bar{y}_B)$ and the sample covariance matrix

$$S = \begin{pmatrix} S_{AA} & S_{AB} \\ S_{BA} & S_{BB} \end{pmatrix}$$

have been partitioned like μ to correspond with the drug phases. The numbers come out as $\bar{y}_A = (0.928, 1.62, 1.18, 0.840)$, $\bar{y}_B = (0.910, 1.58, 1.37, 0.982)$,

$$S_{AA} = \begin{pmatrix} 0.427 & \cdot & \cdot & \cdot \\ 0.672 & 1.11 & \cdot & \cdot \\ 0.288 & 0.471 & 0.264 & \cdot \\ 0.302 & 0.530 & 0.268 & 0.310 \end{pmatrix}$$

$$S_{AB} = \begin{pmatrix} -0.087 & 0.019 & 0.193 & 0.176 \\ -0.050 & 0.128 & 0.444 & 0.388 \\ -0.013 & 0.390 & 0.527 & 0.429 \\ 0.114 & 0.346 & 0.550 & 0.445 \end{pmatrix},$$

$$S_{BB} = \begin{pmatrix} 1.15 & \cdot & \cdot & \cdot \\ 0.037 & 2.09 & \cdot & \cdot \\ 0.507 & 2.12 & 2.44 & \cdot \\ 0.333 & 1.67 & 1.91 & 1.50 \end{pmatrix}$$

so $T^2 = n(\bar{y}_A - \bar{y}_B)^T(S_{AA} - 2S_{AB} + S_{BB})^{-1}(\bar{y}_A - \bar{y}_B) = 6.75$ and $F_{4,1} = 0.42$.

This is only a pilot study with $n = 5$ and the denominator degrees of freedom are much too few for useful inference. However, the technique is illustrated. Notice that, although S is singular because $n - 1 < p$, it is not S but HSH^T which has to be inverted for T^2, and this is nonsingular. The formula given for T^2 above, involving $\bar{y}_A - \bar{y}_B$, is recognizable as the multivariate extension of the familiar paired t-test to which it reduces in the case $p = 2$. This form occurs in biometry where the 'phases' comprise corresponding measurements made on the left and right sides of an animal or plant and the hypothesis to be tested is of symmetry. Extensions to this sort of thing are obvious. □

Example 4.4

The Anova for Example 3.1, involving three groups of guinea pigs, did not establish differences between groups in either overall level of the growth profiles or their shape. Some tests can also be made using Hotelling's T^2 as follows.

First, the T^2 value for a difference between the mean vectors of Groups 1 and 2 is $T_{12}^2 = 32.0$ and so $F_{6,7} = 3.11$ ($0.05 < p < 0.10$). Similarly, for the other group-pairs $T_{13}^2 = 29.7$ ($F_{6,7} = 2.89$, $0.05 < p < 0.10$) and $T_{23}^2 = 24.4$ ($F_{6,7} = 2.38$, $p > 0.10$). The covariance matrix pooled from all three groups, with 12 degrees of freedom, was used for S in computing these T^2 values. These tests do not yield strong evidence of differences.

The corresponding T^2 tests for parallelism of group profiles yield $T_{12}^2 = 31.9$ ($F_{5,8} = 4.25$, $0.025 < p < 0.05$), $T_{13}^2 = 14.2$ ($F_{5,8} = 1.90$, $p > 0.10$) and $T_{23}^2 = 11.3$ ($F_{5,8} = 1.51$, $p > 0.10$). There is some evidence here of non-parallelism in the test between Groups 1 and 2. ☐

Example 4.5

There are two main effects of interest in the three-way factorial design of Example 3.2. To test differences between eyes T^2 can be applied in exactly the same way as in Example 4.3, with $\boldsymbol{\mu}_A$ and $\boldsymbol{\mu}_B$ corresponding now to left and right eyes. Thus $T^2 = 2.50$ and $F_{4,3} = 0.31$. To test between lenses the appropriate hypothesis matrix is

$$\mathbf{H} = \begin{pmatrix} 1 & -1 & 0 & 0 & 0 & 0 & 0 & 0 \\ 1 & 0 & -1 & 0 & 0 & 0 & 0 & 0 \\ 1 & 0 & 0 & -1 & 0 & 0 & 0 & 0 \\ 0 & 0 & 0 & 0 & 1 & -1 & 0 & 0 \\ 0 & 0 & 0 & 0 & 1 & 0 & -1 & 0 \\ 0 & 0 & 0 & 0 & 1 & 0 & 0 & -1 \end{pmatrix}$$

In this case $T^2 = 98.2$ and $F_{6,1} = 2.73$ ($p > 0.10$); again, this can only be an illustration of the method with such a small sample. ☐

4.3 Testing for polynomial trends

There is a tradition in statistics of fitting polynomials to observed response functions when there is no theoretical basis for the true relationship. The justification for this practice is not always clear-cut, and the case for support often primarily rests on simplicity, both in the description of the curve and in the fitting of the associated linear model; smoothing and summarizing the data are often mentioned in this context. Optimistically, a low-degree polynomial will suffice, the interpretation of anything beyond cubic often straining credibility. However, if a polynomial is to be fitted then one could do a lot worse than apply the following neat approach to testing goodness-of-fit due to Hills (1968).

Suppose first that we wish to test whether the mean profile $(\mu_1, \ldots, \mu_p)$ follows a straight line, the corresponding x-points being $(x_1, \ldots, x_p)$. Consider just the first three mean levels (μ_1, μ_2, μ_3) at x-values (x_1, x_2, x_3). The three points lie on a straight line if the gradient between (x_1, μ_1) and (x_2, μ_2) is equal to that between (x_2, μ_2) and (x_3, μ_3), i.e. if $(\mu_2 - \mu_1)/(x_2 - x_1) = (\mu_3 - \mu_2)/(x_3 - x_2)$. The quantities appearing on either side of the equation here are called **divided differences**. With equi-spaced x's we could work in terms of undivided differences, $\mu_2 - \mu_1$ and $\mu_3 - \mu_2$. In matrix terms the condition is $\mathbf{H}\mathbf{\mu} = \mathbf{0}$ where $\mathbf{\mu} = (\mu_1, \mu_2, \mu_3)^T$, $\mathbf{H}(1 \times 3) = (-d_1, d_1 + d_2, -d_2)$ and $d_j = (x_{j+1} - x_j)^{-1}$ for $j = 1, 2$. Extending this now to the case $p > 3$, we take $\mathbf{H}$ to be a $(p - 2) \times p$ matrix:

$$\mathbf{H} = \begin{pmatrix} -d_1 & d_1 + d_2 & -d_2 & 0 & 0 & \cdots & 0 \\ 0 & -d_2 & d_2 + d_3 & -d_3 & 0 & \cdots & 0 \\ 0 & 0 & -d_3 & d_3 + d_4 & -d_4 & \cdots & 0 \\ \cdot & \cdot & \cdot & \cdot & \cdot & \cdots & \cdot \end{pmatrix} \quad (4.3.1)$$

The straight-line hypothesis is that $\mathbf{H}\mathbf{\mu} = \mathbf{0}$. Hotelling's T^2 can now be applied with $\mathbf{z} = n^{1/2}\mathbf{H}\bar{\mathbf{y}}$ and $\mathbf{A} = v\mathbf{H}\mathbf{S}\mathbf{H}^T$ where n and $\bar{\mathbf{y}}$ are the sample size and mean, and $\mathbf{S}$ is the sample covariance matrix with $v = n - 1$ degrees of freedom.

The approach can be generalized to quadratic curves, cubics and higher-order polynomials as follows. Define the **forward shift operator** F by $Fx_j = x_{j+1}$; thus, e.g. $(F-1)x_j = x_{j+1} - x_j$ and $F^2\mu_j = \mu_{j+2}$. The divided differences used above are then $(\mu_2 - \mu_1)/(x_2 - x_1) = (F-1)\mu_1/(F-1)x_1$, and generally $(F-1)\mu_j/(F-1)x_j \, (j = 1, \ldots, p-1)$. Let us denote these first divided differences by D_{1j}. Then second divided differences are defined as $D_{2j} = (F-1)D_{1j}/(F^2 - 1)x_j$; for example,

$$\begin{aligned} D_{21} &= (D_{12} - D_{11})/(x_3 - x_1) \\ &= \{(\mu_3 - \mu_2)/(x_3 - x_2) - (\mu_2 - \mu_1)/(x_2 - x_1)\} \div (x_3 - x_1) \\ &= \mu_1/\{(x_1 - x_2)(x_1 - x_3)\} + \mu_2/\{(x_2 - x_1)(x_2 - x_3)\} \\ &\quad + \mu_3/\{(x_3 - x_1)(x_3 - x_2)\}. \end{aligned} \quad (4.3.2)$$

Continuing along this track, the lth divided differences are $D_{lj} = (F-1)D_{l-1,j}/(F^l - 1)x_j$ $(j = 1, \ldots, p-l)$, and we run out of track when l reaches p. Now, inspection of this process reveals that D_{lj} is a linear combination of μ's, namely μ_j to μ_{j+l}, with coefficients dependent on the x's. In fact (e.g. Hartree, 1958, section 5.72),

$$D_{lj} = \sum_{k=j}^{j+l} h_{ljk}\mu_k \quad \text{with} \quad h_{ljk} = \prod_{s=j}^{j+l}{}'(x_k - x_s)^{-1},$$

the dash in $\prod'$ indicating that the factor with $s = k$ is omitted; (4.3.2) is an example, with $l = 2$ and $j = 1$. It can also be shown that if $\mu_j = f_l(x_j)$, $f_l(\cdot)$ being a polynomial function of order at most l, then $D_{sj} = 0$ for $s > l$. Hence, the H-matrix for testing for an lth-order polynomial trend in the μ_j is constructed as $(p-l) \times p$ with entries: $H_{jk} = h_{ljk}$ for $k = j, \ldots, j+l$; $H_{jk} = 0$ otherwise. The matrix of (4.3.1) can be obtained in this way for $l = 1$, after a harmless rescaling of the rows.

Example 4.4 (contd)

Let us consider whether the growth curves here can be adequately described by straight lines. A separate test for each group will be made, taking n and y in the above account as each group size and mean in turn, but with S as the pooled sample covariance matrix with 12 degrees of freedom. The results are as follows. For Group 1, $T_1^2 = 35.3$ ($F_{4,9} = 6.61$, $0.001 < p < 0.01$); for Group 2, $T_2^2 = 18.8$ ($F_{4,9} = 3.52$, $0.05 < p < 0.10$); and for Group 3, $T_3^2 = 20.3$ ($F_{4,9} = 3.81$, $0.025 < p < 0.05$). The evidence for linearity seems somewhat underwhelming. In similar fashion, that for quadratics ($F_{3,10} = 8.10$, 4.89, 5.03) and cubics ($F_{2,11} = 12.5$, 1.82, 8.29) is equally unimpressive. The guinea pigs stoutly refuse to submit to our favourite curves. □

4.4 Manova

The acronym 'Manova' stands for 'multivariate analysis of variance'. As the name implies, it is just the extension of Anova to the multivariate case of vector observations. With multivariate methods we have seen that univariate variances are replaced by multivariate covariance matrices. In Manova the sums of squares of Anova are replaced by the corresponding s.s.p. (sums of squares and products) matrices. Thus, one can construct an Anova-type table in which the former s.s. column is replaced by one containing s.s.p. matrices corresponding to the various effects. The effects referred to here are only fixed since any random effects have been subsumed into the covariance structure. In Anova the F-tests are based on ratios of the sums of squares, but there is no such unique way of comparing matrices and various aspects have been examined in the literature. These are based on the eigenvalues of the matrices involved in one way or another. Rather than trying to give a comprehensive treatment we will illustrate the method with one or two short examples.

Example 4.1 (contd)

In section 2.1 the data were treated to univariate analyses, concentrating on the observations at time 5 minutes. Thus, Groups 1 and 2 were compared by a t-test, and then this was extended to a comparison of all four groups in a one-way Anova. In section 4.2 Group 1 and 2 were again compared, this time by a T^2-test based on times 1, 5 and 10 minutes. The logical progression is now to perform a one-way Manova between the four groups based on these three times.

The sample sizes for the four groups are $n_1 = 8$, $n_2 = 6$, $n_3 = 7$, $n_4 = 6$ and the mean vectors are

$$\bar{\mathbf{y}}_1 = \begin{pmatrix} 6\cdot83 \\ 6\cdot81 \\ 6\cdot39 \end{pmatrix}, \quad \bar{\mathbf{y}}_2 = \begin{pmatrix} 7\cdot73 \\ 6\cdot98 \\ 8\cdot35 \end{pmatrix}, \quad \bar{\mathbf{y}}_3 = \begin{pmatrix} 5\cdot20 \\ 5\cdot21 \\ 6\cdot00 \end{pmatrix}, \quad \bar{\mathbf{y}}_4 = \begin{pmatrix} 5\cdot43 \\ 5\cdot45 \\ 5\cdot35 \end{pmatrix}$$

The s.s.p. matrix for Group 1 is $(n_1 - 1)\mathbf{S}_1$, where $\mathbf{S}_1$ appears in section 4.2, and similarly for the other groups. Their sum, $\mathbf{T}_W = \sum_g (n_g - 1)\mathbf{S}_g$, is referred to as the **within-groups s.s.p. matrix**. The **total s.s.p. matrix** $\mathbf{T}$ is computed as $(n-1)\mathbf{S}$ where $n = n_1 + n_2 + n_3 + n_4$ and $\mathbf{S}$ is the sample covariance matrix for all the data, ignoring the grouping. The difference, $\mathbf{T} - \mathbf{T}_W = \mathbf{T}_B$, is the **between-groups s.s.p. matrix**; it can be shown that $\mathbf{T}_B = \sum_g n_g (\bar{\mathbf{y}}_g - \bar{\mathbf{y}})(\bar{\mathbf{y}}_g - \bar{\mathbf{y}})^T$ where $\bar{\mathbf{y}} = n^{-1} \sum n_g \bar{\mathbf{y}}_g$ is the overall mean.

The numerical results are

$$\mathbf{T_W} = \begin{pmatrix} 284 & \cdot & \cdot \\ 290 & 326 & \cdot \\ 339 & 376 & 491 \end{pmatrix}, \quad \mathbf{T_B} = \begin{pmatrix} 28 & \cdot & \cdot \\ 21 & 17 & \cdot \\ 25 & 17 & 30 \end{pmatrix}$$

and the analysis is presented in the following table.

One-way Manova for Example 4.6

Source of variation	d.f.	s.s.p matrix
Between groups	3	$\mathbf{T_B}$
Within groups	23	$\mathbf{T_W}$
Total	26	$\mathbf{T}$

As in univariate Anova, strong differences between groups will be established if $\mathbf{T_B}$ is much larger than $\mathbf{T_W}$, or equivalently if $\mathbf{T_W}$ makes a much smaller contribution to $\mathbf{T}$ than does $\mathbf{T_B}$. Wilks's lambda is defined as $\Lambda = \det \mathbf{T_W}/\det \mathbf{T}$ and is equivalent to the likelihood ratio test of the hypothesis of equal group means. Its value here is 0.682 and to decide whether this is small enough to be significant it can be transformed approximately to an F-statistic as $F_{9,51} = (\Lambda^{-1/2.4} - 1) \times 51/9 = 0.97$ (for details see Rao, section 8c.5). $\square$

Example 4.4 (contd)

This experiment provides another one-way layout with three groups, $n_1 = n_2 = n_3 = 5$ and $p = 6$. The resulting Wilks's $\Lambda = 0.088$, and this may be transformed to $F_{12,14} = 2.77$ ($0.025 < p < 0.05$). $\square$

Wilks's lambda is equal to the product of the eigenvalues of $\mathbf{T_W T^{-1}} = (\mathbf{I}_p + \mathbf{T_B T_W^{-1}})^{-1}$. Other test criteria include Roy's largest root (the largest eigenvalue of $\mathbf{T_B T_W^{-1}}$, or alternatively of $\mathbf{T_B T^{-1}}$), the Lawley–Hotelling trace (trace $\mathbf{T_B T_W^{-1}}$, equal to the sum of the eigenvalues of $\mathbf{T_B T_W^{-1}}$) and Pillai's trace (trace $\mathbf{T_W T^{-1}}$). More detailed discussion of these criteria is given by Morrison (1976, section 5.8), and Kenward (1979) gives a nice geometrical interpretation of them in terms of the associated concentration ellipsoids.

4.5 Further reading

There are now many books which cover the general topics of this chapter, though not necessarily with repeated measures in mind. Hand and Taylor (1987), and Chapters 4 and 5 of Morrison (1976) are very relevant. Other accounts are given in Tatsuoka (1971, Chapter 7), Winer (1971, section 3.15), Harris (1975, Chapters 3 and 4), and, for some theory, Rao (1965, Chapter 8). Some discussion of the relative merits, regarding power and robustness, of the various multivariate criteria is given by Olsen (1976).

CHAPTER 5

Regression models

5.1 Special case

We shall investigate, after Rao (1959), the general linear model in which the observed vector $\mathbf{y}$ is p-variate normal with mean $\boldsymbol{\mu} = \mathbf{X}\boldsymbol{\beta}$ and covariance matrix $\boldsymbol{\Sigma}$. Here $\mathbf{X}$ is a $p \times q$ matrix of known explanatory variables or design indicators, $\boldsymbol{\beta}$ is a $q \times 1$ vector of unknown regression coefficients, and $\boldsymbol{\Sigma}(p \times p)$ is unknown and unstructured. Suppose that one has a sample of n such vector observations $\mathbf{y}_1, \ldots, \mathbf{y}_n$ from $N_p(\mathbf{X}\boldsymbol{\beta}, \boldsymbol{\Sigma})$. The special case referred to in the title of this section is that the $\mathbf{y}_i$ all have the same $\mathbf{X}$ matrix. Let the sample mean vector be $\bar{\mathbf{y}}$ and the sample covariance matrix be $\mathbf{S}$. When $\boldsymbol{\Sigma}$ is known the generalized least-squares estimator $\bar{\boldsymbol{\beta}}$ is found by minimizing the quadratic form $\sum_{i=1}^{n} Q_i(\boldsymbol{\beta}, \boldsymbol{\Sigma})$ where

$$Q_i(\boldsymbol{\beta}, \boldsymbol{\Sigma}) = (\mathbf{y}_i - \mathbf{X}\boldsymbol{\beta})^{\mathrm{T}} \boldsymbol{\Sigma}^{-1} (\mathbf{y}_i - \mathbf{X}\boldsymbol{\beta}); \tag{5.1.1}$$

The solution is $\bar{\boldsymbol{\beta}} = (\mathbf{X}^{\mathrm{T}} \boldsymbol{\Sigma}^{-1} \mathbf{X})^{-1} \mathbf{X}^{\mathrm{T}} \boldsymbol{\Sigma}^{-1} \bar{\mathbf{y}}$. The covariance matrix of $\bar{\boldsymbol{\beta}}$ is $\mathbf{V}_\beta = n^{-1} (\mathbf{X}^{\mathrm{T}} \boldsymbol{\Sigma}^{-1} \mathbf{X})^{-1}$ from which standard errors may be calculated for individual coefficients and tests may be made for contrasts, among other things. When $\boldsymbol{\Sigma}$ is unknown, and this is usually the case in practice, it is replaced by its estimator $\mathbf{S}$, so then $\hat{\boldsymbol{\beta}} = (\mathbf{X}^{\mathrm{T}} \mathbf{S}^{-1} \mathbf{X})^{-1} \mathbf{X}^{\mathrm{T}} \mathbf{S}^{-1} \bar{\mathbf{y}}$. Under the normal distribution assumption $\hat{\boldsymbol{\beta}}$ is also the maximum likelihood estimator. The covariance matrix of $\hat{\boldsymbol{\beta}}$ is likewise estimated as $\hat{\mathbf{V}}_\beta = n^{-1}(\mathbf{X}^{\mathrm{T}} \mathbf{S}^{-1} \mathbf{X})^{-1}$.

For testing the goodness-of-fit of the model $\boldsymbol{\mu} = \mathbf{X}\boldsymbol{\beta}$ we argue as follows, using the familiar extra-sum-of-squares argument here applied to quadratic forms. The residual form with $\boldsymbol{\mu}$ unconstrained, and hence estimated by $\bar{\mathbf{y}}$, is $\sum_{i=1}^{n} (\mathbf{y}_i - \bar{\mathbf{y}}) \mathbf{S}^{-1} (\mathbf{y}_i - \bar{\mathbf{y}})$. The residual with $\boldsymbol{\mu}$ constrained to be of the form $\mathbf{X}\boldsymbol{\beta}$, and hence estimated by $\mathbf{X}\hat{\boldsymbol{\beta}}$, is $\sum_{i=1}^{n} (\mathbf{y}_i - \mathbf{X}\hat{\boldsymbol{\beta}})^{\mathrm{T}} \mathbf{S}^{-1} (\mathbf{y}_i - \mathbf{X}\hat{\boldsymbol{\beta}})$. The difference between these two, which represents how much better $\bar{\mathbf{y}}$ fits the data than $\mathbf{X}\hat{\boldsymbol{\beta}}$, can be manipulated algebraically into the form $n \bar{\mathbf{y}}^{\mathrm{T}} \mathbf{S}^{-1} (\mathbf{I}_p - \mathbf{P}) \bar{\mathbf{y}}$, where

I_p is the $p \times p$ unit matrix and $P = X(X^T S^{-1} X)^{-1} X^T S^{-1}$. This difference can be converted to a variance ratio statistic, and the resulting goodness-of-fit test is based on

$$F_{p-q, n-p+q} = (n - p + q)(p - q)^{-1} \bar{y}^T S^{-1} (I_p - P) \bar{y}.$$

Rao (1959) gives two alternative approaches to the goodness-of-fit test, the second of which is similar to that just presented. It is based on minimizing the quadratic form $Q(\beta) = (\bar{y} - X\beta)^T S^{-1} (\bar{y} - X\beta)$ over β; this corresponds to choosing β so that $X^T \beta$ is close to $\bar{y}$. The minimum is achieved at $\hat{\beta}$, given above, and $Q(\hat{\beta}) = \bar{y}^T S^{-1} (I_p - P) \bar{y}$, the same form as before.

Example 5.1

Twelve patients were observed during three phases $(2 + 3 + 2$ occasions). The model assumed in the Anova of Example 3.3, and tested by Hotelling's T^2 in Example 4.2, implies constant mean values within phases. In regression terms the mean vector $\mu(7 \times 1)$ is expressed as $X\beta$ where $\beta = (\alpha_1, \alpha_2, \alpha_3)^T$ comprises the three phase-means and X is 7×3:

$$X^T = \begin{pmatrix} 1 & 1 & 0 & 0 & 0 & 0 & 0 \\ 0 & 0 & 1 & 1 & 1 & 0 & 0 \\ 0 & 0 & 0 & 0 & 0 & 1 & 1 \end{pmatrix}.$$

The computed values, with $n = 12$, $p = 7$ and $q = 3$, are

$$\hat{\beta} = \begin{pmatrix} 0.742 \\ 1.15 \\ 0.776 \end{pmatrix}, \quad \hat{V}_\beta = 10^{-3} \begin{pmatrix} 4.16 & \cdot & \cdot \\ 2.76 & 2.41 & \cdot \\ 0.212 & 0.111 & 2.58 \end{pmatrix}$$

and $F_{4,8} = 2.46$ $(p > 0.10)$ for goodness of fit. Actually, this F-test is the same as that applied in Example 4.2, the difference being attributed to inaccuracies in the computations. (The experts will note that (H^T, X) has full rank and $HX = 0$, with H given in Example 4.2.) That the mean level in phase two is different from the others is pretty obvious on inspection of the estimates and their standard errors. For instance $\hat{\alpha}_2 - \hat{\alpha}_1 = 0.41$ with estimated variance $10^{-3}(4.16 - 2 \times 2.76 + 2.41) = 0.00105$ and standard error 0.03. However, $\hat{\alpha}_3 - \hat{\alpha}_1 = 0.034$ with standard error 0.079. $\square$

5.2 General case

The analysis presented in the previous section is restricted to the case where the observations y_i $(i = 1, \ldots, n)$ all have the same $N_p(X\beta, \Sigma)$ distribution. We now discuss the case where the X's may differ between individuals, so that cases where the individuals are in different treatment groups or have different covariates are now covered. Further, missing values will be allowed for by not requiring the y_i to be all of full length p. The y_i will now be independently distributed as $N_{p_i}(X_i\beta, \Sigma_i)$ where X_i is $p_i \times q$ and Σ_i is the $p_i \times p_i$ submatrix of the full Σ corresponding to the observed components of y_i. Thus, y_i has $p - p_i$ missing components (none if $p_i = p$) and Σ_i is obtained from Σ by deleting the rows and columns corresponding to the missing components; throughout this section, Σ is unknown and unstructured.

The generalized least-squares estimator $\hat{\beta}$ is found by minimizing the quadratic form $\sum_{i=1}^{n} Q_i(\beta, \Sigma)$ where

$$Q_i(\beta, \Sigma) = (y_i - X_i\beta)^T \Sigma_i^{-1} (y_i - X_i\beta) \qquad (5.2.1)$$

in the hypothetical case of known Σ. The solution is

$$\hat{\beta} = \left(\sum_{i=1}^{n} X_i^T \Sigma_i^{-1} X_i \right)^{-1} \left(\sum_{i=1}^{n} X_i^T \Sigma_i^{-1} y_i \right) \qquad (5.2.2)$$

with covariance matrix $V_\beta = (\sum_{i=1}^{n} X_i^T \Sigma_i^{-1} X_i)^{-1}$. When Σ is unknown the maximum likelihood estimator (m.l.e.) of β has the same form but with Σ_i replaced by $\hat{\Sigma}_i$, the appropriate submatrix of the m.l.e. $\hat{\Sigma}$. When there are no missing values $\hat{\Sigma}$ is given by $\hat{\Sigma} = n^{-1} \sum_{i=1}^{n} (y_i - X_i\hat{\beta})(y_i - X_i\hat{\beta})^T$. When there are, the equations are intractable and a simple course is to retain this basic form for $\hat{\Sigma}$, i.e. with $\hat{\Sigma}_{jk}$ being the average of $(y_i - X_i\hat{\beta})_j (y_i - X_i\hat{\beta})_k$ over the sample, but averaging only over the components present for each (j, k) combination. The divisor $n - q$ is often preferred to n for estimating Σ; this takes account of the loss of degrees of freedom in estimating β and yields unbiased estimates in balanced cases. In general, there is no explicit solution for $\hat{\beta}$ and $\hat{\Sigma}$ separately and the equations have to be solved iteratively. For example, a see-saw method starting with S (the overall sample covariance matrix) for $\hat{\Sigma}$, then solving the first equation for $\hat{\beta}$, and then the second for $\hat{\Sigma}$, and so on back and forth, often converges in a few cycles. Failing this, the general method of section 5.3 is applicable. There are closed-form solutions in particular situations but we shall not investigate these special cases. Also, unfortunately, the F-test for goodness-of-fit given in section 5.1 is no longer available.

The estimated covariance matrix of $\hat{\boldsymbol{\beta}}$ is $\hat{\mathbf{V}}_\beta = (\sum_{i=1}^n \mathbf{X}_i^T \hat{\boldsymbol{\Sigma}}_i^{-1} \mathbf{X}_i)^{-1}$ and this may be used for hypothesis tests and confidence intervals in the usual way. For instance, for the linear hypothesis $\mathbf{H}\boldsymbol{\beta} = \mathbf{h}$, where $\mathbf{H}(r \times q)$ and $\mathbf{h}(r \times 1)$ are specified, the natural statistic to use is $\mathbf{T}_H = (\mathbf{H}\hat{\boldsymbol{\beta}} - \mathbf{h})^T (\mathbf{H}\hat{\mathbf{V}}_\beta \mathbf{H}^T)^{-1} (\mathbf{H}\hat{\boldsymbol{\beta}} - \mathbf{h})$. This has an approximate χ_r^2 distribution in large samples. The reason for this is that, as $n \to \infty$, $\hat{\boldsymbol{\Sigma}} \underset{P}{\to} \boldsymbol{\Sigma}$ (in probability) and $\hat{\boldsymbol{\beta}} \underset{D}{\to} N_q(\boldsymbol{\beta}, \mathbf{v}_\beta)$ (in distribution). Thus, under the hypothesis, $\mathbf{H}\hat{\boldsymbol{\beta}} - \mathbf{h} \underset{D}{\to} N_r(\mathbf{0}, \mathbf{H}\mathbf{V}_\beta \mathbf{H}^T)$ and so $\mathbf{T}_H \underset{D}{\to} \chi_r^2$.

Example 5.2

The data, with previous much acclaimed appearances in Examples 3.1 and 4.4, comprise growth profiles for three groups of guinea pigs. The results to date indicate little evidence of differences between mean group profiles, and little support for low-degree polynomial growth curves. A single straight-line regression fitted to all the data yields $\hat{\boldsymbol{\beta}} = (488, 26.1)$ and $F_{4,11} = 8.20$ $(0.001 < p < 0.005)$ for goodness of fit in accordance with the previous tests. Likewise, a single cubic fit yields $F_{2,13} = 14.0$ $(p < 0.001)$.

Purely for illustration of the current technique let us fit parallel growth lines $\mu_j = \alpha_g + \delta x_j$ with intercepts $\alpha_1, \alpha_2, \alpha_3$ for the three groups and common slope δ. In the notation of this section $p = 6$, $q = 4$, $\boldsymbol{\beta} = (\alpha_1, \alpha_2, \alpha_3, \delta)^T$ and $\mathbf{X}$ has jth row $(1, 0, 0, x_j)$ for Group 1, $(0, 1, 0, x_j)$ for Group 2, and $(0, 0, 1, x_j)$ for Group 3. The results, after ten cycles of the iteration described above, are

$$\hat{\boldsymbol{\beta}} = \begin{pmatrix} 478 \\ 481 \\ 513 \\ 26.1 \end{pmatrix}, \quad \hat{\mathbf{V}}_\beta = \begin{pmatrix} 47.9 & \cdot & \cdot & \cdot \\ 15.6 & 47.9 & \cdot & \cdot \\ 15.6 & 15.6 & 47.9 & \cdot \\ -6.22 & -6.22 & -6.22 & 2.48 \end{pmatrix}.$$

The mean profile levels look equal for Groups 1 and 2 but that for Group 3 looks different: $\hat{\alpha}_3 - \hat{\alpha}_1 = 35.0$ with estimated variance $47.9 - 2 \times 15.6 + 47.9 = 64.6$ and standard error 8.04. □

Example 5.3

The effect of a test protein diet on the growth of chicks over their first three weeks was investigated by a nutrition student. There were

Table 5.1. *Growth of chicks under different protein diets*

Chick	Day 0	2	4	6	8	10	12	14	16	18	20	21
Group 1 1	42	51	59	64	76	93	106	125	149	171	199	205
2	40	49	58	72	84	103	122	138	162	187	209	215
3	43	39	55	67	84	99	115	138	163	187	198	202
4	42	49	56	67	74	87	102	108	136	154	160	157
5	41	42	48	60	79	106	141	164	197	199	220	223
6	41	49	59	74	97	124	141	148	155	160	160	157
7	41	49	57	71	89	112	146	174	218	250	288	305
8	42	50	61	71	84	93	110	116	126	134	125	-1
9	42	51	59	68	85	96	90	92	93	100	100	98
10	41	44	52	63	74	81	89	96	101	112	120	124
11	43	51	63	84	112	139	168	177	182	184	181	175
12	41	49	56	62	72	88	119	135	162	185	195	205
13	41	48	53	60	65	67	71	70	71	81	91	96
14	41	49	62	79	101	128	164	192	227	248	259	266
15	41	49	56	64	68	68	67	68	−1	−1	−1	−1
16	41	45	49	51	57	51	54	−1	−1	−1	−1	−1
17	42	51	61	72	83	89	98	103	113	123	133	142
18	39	35	−1	−1	−1	−1	−1	−1	−1	−1	−1	−1
19	43	48	55	62	65	71	82	88	106	120	144	157
20	41	47	54	58	65	73	77	89	98	107	115	117
Group 2 21	40	50	62	86	125	163	217	240	275	307	318	331
22	41	55	64	77	90	95	108	111	131	148	164	167
23	43	52	61	73	90	103	127	135	145	163	170	175
24	42	52	58	74	66	68	70	71	72	72	76	74
25	40	49	62	78	102	124	146	164	197	231	259	265
26	42	48	57	74	93	114	136	147	169	205	236	251
27	39	46	58	73	87	100	115	123	144	163	185	192
28	39	46	58	73	92	114	145	156	184	207	212	233
29	39	48	59	74	87	106	134	150	187	230	279	309
30	42	48	59	72	85	98	115	122	143	151	157	150
Group 3 31	42	53	62	73	85	102	123	138	170	204	235	256
32	41	49	65	82	107	129	159	179	221	263	291	305
33	39	50	63	77	96	111	137	144	151	146	156	147
34	41	49	63	85	107	134	164	186	235	294	327	341
35	41	53	64	87	123	158	201	238	287	332	361	373
36	39	48	61	76	98	116	145	166	198	227	225	220
37	41	48	56	68	80	83	103	112	135	157	169	178
38	41	49	61	74	98	109	128	154	192	232	280	290
39	42	50	61	78	89	109	130	146	170	214	250	272
40	41	55	66	79	101	120	154	182	215	262	295	321
Group 4 41	42	51	66	85	103	124	155	153	175	184	199	204
42	42	49	63	84	103	126	160	174	204	234	269	281
43	42	55	69	96	131	157	184	188	197	198	199	200
44	42	51	65	86	103	118	127	138	145	146	−1	−1
45	41	50	61	78	98	117	135	141	147	174	197	196
46	40	52	62	82	101	120	144	156	173	210	231	238
47	41	53	66	79	100	123	148	157	168	185	210	205
48	39	50	62	80	104	125	154	170	222	261	303	322
49	40	53	64	85	108	128	152	166	184	203	233	237
50	41	54	67	84	105	122	155	175	205	234	264	264

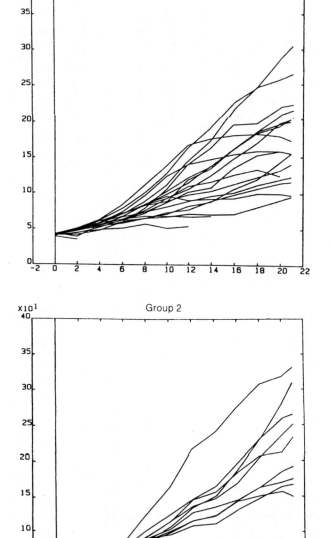

Fig. 5.1. *Growth of chicks under different protein diets*

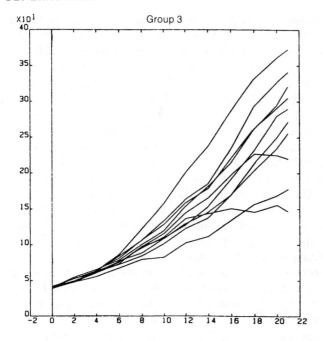

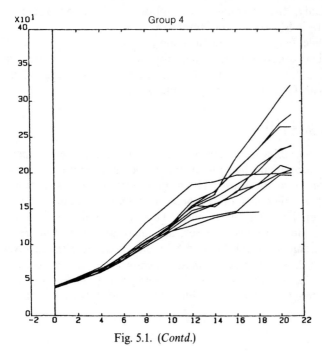

Fig. 5.1. (*Contd.*)

four groups of chicks: control (normal diet) and three test groups (10%, 20% and 40% protein replacement). The body weights were recorded on alternate days, and the questions concern differences in growth rates between groups. The data are given in Table 5.1 and the corresponding pictures in Fig. 5.1. These are classic growth curves showing the typical fan pattern as individual lines diverge.

Inspection of the plots suggests a quadratic curve so, allowing for different intercepts between groups, the model $\mu_j = \alpha_g + \beta_1 x_j + \beta_2 x_j^2$ will be tried. The five rows with missing observations correspond to animals that had to be withdrawn during the study. They will be omitted here so that inferences will be made about healthy animals only. The numerical estimates turn out as $(\hat{\alpha}_1, \hat{\alpha}_2, \hat{\alpha}_3, \hat{\alpha}_4) = (41\cdot7, 40\cdot9, 41\cdot4, 41\cdot3)$ and $(\hat{\beta}_1, \hat{\beta}_2) = (3\cdot26, 0\cdot111)$; the standard errors of the $\hat{\alpha}$'s are all around $0\cdot3$, and those of $\hat{\beta}_1$ and $\hat{\beta}_2$ are $0\cdot15$ and $0\cdot021$. Thus the intercepts do not look very different, and the quadratic coefficient β_2 is probably nonzero.

However, the fit of the quadratic model is suspect. Proceeding as in section 4.3, the T^2 tests based on differencing yield $F_{9,33} = 2\cdot15$, $7\cdot84$, $8\cdot29$ and $17\cdot2$ respectively for the four groups. Only Group 1 seems cooperative, and the results are similarly unrewarding for all differences up to seventh order (testing a polynomial of degree 6). Certainly, the attempted description of the growth curves as quadratic is shown to be simplistic.

Incidentally, to address the main purpose of the investigation, T^2 tests between the groups, as described in section 4.2, were performed. These indicated equality of the mean profiles in Groups 2, 3 and 4, but with Group 1 (control) in splendid isolation.   □

Example 5.4

Many variables were recorded on $n = 30$ hip-replacement patients, the measurements being made on one occasion before the operation and on three afterwards. The primary interest in this study was in possible differences in response between men and women, and whether there was an age effect. Thus, true to life, the patients are broken down by age and sex, and there are up to four repeated measurements. Data for one variable (haematocrit) are given in Table 5.2, where there are many missing values indicated as $-1\cdot0$. Figure 5.2 displays the individual time profiles. These show a fairly consistent trend, though at widely differing levels between patients.

Table 5.2. *Recovery of hip-replacement patients*

Patient		Age	Occasions 1	2	3	4
Male	1	66	47.10	31.05	−1.00	32.80
	2	70	44.10	31.50	−1.00	37.00
	3	44	39.70	33.70	−1.00	24.50
	4	70	43.30	18.35	−1.00	36.60
	5	74	37.40	32.25	−1.00	29.05
	6	65	45.70	35.50	−1.00	39.80
	7	54	44.90	34.10	−1.00	32.05
	8	63	42.90	32.05	−1.00	−1.00
	9	71	46.05	28.80	−1.00	37.80
	10	68	42.10	34.40	34.00	36.05
	11	69	38.25	29.40	32.85	30.50
	12	64	43.00	33.70	34.10	36.65
	13	70	37.80	26.60	26.70	30.60
Female	14	60	37.25	26.50	−1.00	38.45
	15	52	−1.00	27.95	−1.00	33.95
	16	52	27.00	32.50	−1.00	31.95
	17	75	38.35	32.30	−1.00	37.90
	18	72	38.80	32.55	−1.00	26.85
	19	54	44.65	32.25	−1.00	34.20
	20	71	38.00	27.10	−1.00	37.85
	21	58	34.00	23.20	−1.00	25.95
	22	77	44.80	37.20	−1.00	29.70
	23	66	45.95	29.10	−1.00	26.70
	24	53	41.85	31.95	37.15	37.60
	25	74	38.00	31.65	38.40	35.70
	26	78	42.20	34.00	32.90	33.25
	27	74	39.70	33.45	26.60	32.65
	28	79	37.50	28.20	28.80	30.30
	29	71	34.55	30.95	30.60	28.75
	30	68	35.50	24.70	28.10	29.75

For tidiness we will omit the data for occasion 3 altogether, and fill in the other two missing values as 32·05 (subject 8, occasion 4) and 40·00 (subject 15, occasion 1). Also, the value 18·35 for subject 4 on occasion 2 looks odd, but this will be left as it is; in an extended analysis one would need to consider the question of possible outliers, robustness and influence.

In the regression model, with $p = 3$ for occasions 1, 2 and 4, take

$$\mathbf{X}_i = \begin{pmatrix} 1 & 0 & 0 & s_i & a_i \\ 0 & 1 & 0 & s_i & a_i \\ 0 & 0 & 1 & s_i & a_i \end{pmatrix}$$

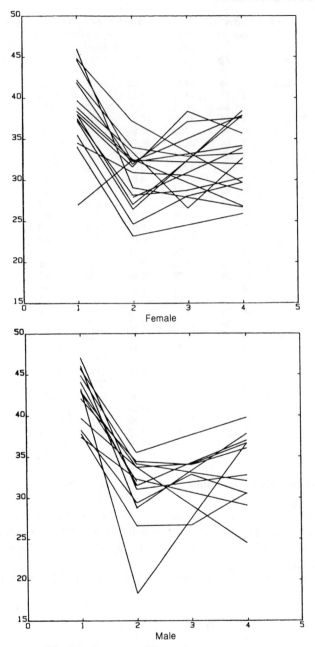

Fig. 5.2. *Recovery of hip-replacement patients.*

where $s_i = 1 \, (-1)$ if the ith patient is male (female), and a_i is his/her age. Hence, the mean value of y_i, modelled as $\mathbf{X}_i\boldsymbol{\beta}$, has components $\beta_j + \beta_4 s_i + \beta_5 a_i$ for $j = 1, 2, 3$ (occasions 1, 2, 4). The only real model restriction here is to allow for the age effect only linearly and with the same coefficient for men and women, i.e. as in traditional analysis of covariance. The estimates, which stabilized after five iterations, are $\hat{\boldsymbol{\beta}}^{\mathrm{T}} = (38{\cdot}4, 28{\cdot}6, 31{\cdot}0, 0{\cdot}807, 0{\cdot}031)$,

$$\hat{\boldsymbol{\Sigma}} = \begin{pmatrix} 16{\cdot}6 & \cdot & \cdot \\ 3{\cdot}19 & 15{\cdot}6 & \cdot \\ 3{\cdot}88 & 0{\cdot}254 & 16{\cdot}5 \end{pmatrix}$$

and the estimated covariance matrix of $\hat{\boldsymbol{\beta}}$ is

$$\hat{\mathbf{V}}_\beta = \begin{pmatrix} 13{\cdot}6 & \cdot & \cdot & \cdot & \cdot \\ 13{\cdot}2 & 13{\cdot}6 & \cdot & \cdot & \cdot \\ 13{\cdot}2 & 13{\cdot}1 & 13{\cdot}6 & \cdot & \cdot \\ -0{\cdot}114 & -0{\cdot}114 & -0{\cdot}114 & 0{\cdot}237 & \cdot \\ -0{\cdot}198 & -0{\cdot}198 & -0{\cdot}198 & 0{\cdot}0022 & 0{\cdot}0030 \end{pmatrix}.$$

The coefficients $\hat{\beta}_4$ and $\hat{\beta}_5$, incorporating the sex and age effects, do not greatly exceed their standard errors. $\square$

5.3 Structured covariance case

We consider now the case where $\boldsymbol{\Sigma}$ has a structure specified in terms of some parameter vector $\boldsymbol{\phi}$. The prototype model is $\boldsymbol{\Sigma} = \sigma^2 \mathbf{I}_p$, when $\boldsymbol{\phi}$ has a single entry σ and the components of $\mathbf{y}$ are uncorrelated each with variance σ^2. A slightly more general model is $\boldsymbol{\Sigma} = \sigma^2 \mathbf{I}_p + \sigma_\alpha^2 \mathbf{J}_p$; then $\boldsymbol{\phi} = (\sigma, \sigma_\alpha)$. This is the compound symmetry form which figured largely in Chapter 3 (see section 3.5 in particular). At the other extreme, the unstructured case of sections 5.1 and 5.2 is recovered if $\boldsymbol{\phi}$ contains all the distinct elements of $\boldsymbol{\Sigma}$. Some commonly assumed forms for $\boldsymbol{\Sigma}$ are listed in section 5.4.

The set-up for this section is that the $\mathbf{y}_i$ $(i = 1, \ldots, n)$ are independent, with distributions $\mathrm{N}_p(\mathbf{X}_i\boldsymbol{\beta}, \boldsymbol{\Sigma}_i)$; $\boldsymbol{\Sigma}_i$ is a specified function of $\boldsymbol{\phi}$ and $\boldsymbol{\phi}$ is homogeneous over individuals. For instance, in the missing data situation $\boldsymbol{\Sigma}_i$ will be an appropriate submatrix of $\boldsymbol{\Sigma}(\boldsymbol{\phi})$ as in section 5.2. If $\boldsymbol{\phi}$ were known then the generalized least-squares estimator for $\boldsymbol{\beta}$ could be obtained by minimizing the quadratic form $\sum_{i=1}^{n} Q_i(\boldsymbol{\beta}, \boldsymbol{\phi})$ as in (5.2.1) and (5.2.2). However, $\boldsymbol{\phi}$ will not generally be known and we shall now outline its maximum likelihood estimation.

The log-likelihood function is

$$L_n(\boldsymbol{\beta}, \boldsymbol{\phi}) = \log \prod_{i=1}^{n} \left\{ [\det(2\pi\boldsymbol{\Sigma}_i)]^{-1/2} \exp\left[-\tfrac{1}{2} Q_i(\boldsymbol{\beta}, \boldsymbol{\phi}) \right] \right\}$$

$$= -\tfrac{1}{2} p_+ \log(2\pi) - \tfrac{1}{2} \sum_{i=1}^{n} \left\{ \log \det \boldsymbol{\Sigma}_i + Q_i(\boldsymbol{\beta}, \boldsymbol{\phi}) \right\}, \quad (5.3.1)$$

where $p_+ = \sum p_i$. Maximum likelihood estimators for $\boldsymbol{\beta}$ and $\boldsymbol{\phi}$ are found by maximizing $L_n(\boldsymbol{\beta}, \boldsymbol{\phi})$. For $\boldsymbol{\beta}$ this corresponds to minimizing $\sum Q_i(\boldsymbol{\beta}, \boldsymbol{\phi})$, so $\hat{\boldsymbol{\beta}}$ of (5.2.2) is the m.l.e. also. This forms the basis of a technique for computing the m.l.e.s. The parameter $\boldsymbol{\beta}$ in (5.3.1) is replaced by $\hat{\boldsymbol{\beta}}$ to yield a function $\hat{L}_n(\boldsymbol{\phi})$ of $\boldsymbol{\phi}$ only. In different contexts $\hat{L}_n(\boldsymbol{\phi})$ is variously known as the **maximized, concentrated, reduced** or **profile log-likelihood**. The advantage is that we now have a function of fewer parameters to maximize. A basic computational scheme is to minimize $-\hat{L}_n(\boldsymbol{\phi})$ using a standard routine for function minimization such as the DFP algorithm (Fletcher and Powell, 1963). Modern implementations of such techniques are available in the NAG library. For this the derivative of $\hat{L}_n(\boldsymbol{\phi})$ is often required:

$$\partial \hat{L}_n(\boldsymbol{\phi})/\partial \boldsymbol{\phi}_j = -\tfrac{1}{2} \sum_{i=1}^{n} \left\{ \operatorname{trace}(\boldsymbol{\Sigma}_i^{-1} \boldsymbol{\Sigma}'_{ij}) + Q'_{ij}(\hat{\boldsymbol{\beta}}, \boldsymbol{\phi}) \right\}$$

where

$$\boldsymbol{\Sigma}'_{ij} = \partial \boldsymbol{\Sigma}_i / \partial \boldsymbol{\phi}_j, \qquad Q'_{ij}(\hat{\boldsymbol{\beta}}, \boldsymbol{\phi}) = \left\{ \partial Q_i(\boldsymbol{\beta}, \boldsymbol{\phi})/\partial \boldsymbol{\phi}_j \right\}_{\boldsymbol{\beta}=\hat{\boldsymbol{\beta}}}.$$

Some authors write $\boldsymbol{\Sigma}$ as $\sigma^2 \boldsymbol{\Phi}$ and 'maximize out' the scale factor σ^2 as well as $\boldsymbol{\beta}$. This gives a further reduction of one in the number of parameters in $\hat{L}_n$.

The parameter vector $\boldsymbol{\phi}$ usually plays the role of nuisance parameter in which there is no direct interest, but only in its effect on inferences about $\boldsymbol{\beta}$. If $\boldsymbol{\phi}$ is studied at all, it is usually just to see whether the covariance structure can be simplified or needs elaboration. This is normally done by likelihood-ratio testing; see the examples in section 6.4. The use of e.g. standard errors and t-tests for components and functions of $\boldsymbol{\phi}$ is less common. This is because the large-sample normal approximation to the distribution of m.l.e.s is less satisfactory when the quantities concerned have restricted ranges, and those connected with $\boldsymbol{\phi}$ usually do (e.g. variances, which are positive).

A vector of residuals may be defined as $\hat{\mathbf{u}}_i = \mathbf{y}_i - \mathbf{X}_i\hat{\boldsymbol{\beta}}$, a standardized version being $\hat{\mathbf{R}}_i\hat{\mathbf{u}}_i$ where $\hat{\mathbf{R}}_i$ is any square root of $\hat{\boldsymbol{\Sigma}}_i$, i.e. such that $\hat{\mathbf{R}}_i\hat{\boldsymbol{\Sigma}}_i\hat{\mathbf{R}}_i^T = \mathbf{I}_{p_i}$. Since $\mathbf{R}_i(\mathbf{y}_i - \mathbf{X}_i\boldsymbol{\beta})$, based on the true parameters $\boldsymbol{\Sigma}_i$ and $\boldsymbol{\beta}$, has distribution $N_{p_i}(\mathbf{0}, \mathbf{I}_{p_i})$, the components of $\hat{\mathbf{R}}_i\hat{\mathbf{u}}_i$ should resemble a set of p_i independent standard normal variates. Hence, these provide a test of fit of the model.

5.4 Some covariance structures

In this section some commonly applied models for $\boldsymbol{\Sigma}$ are described.

5.4.1 Equicorrelated errors

The next step up from independent errors is to equicorrelated errors. In this model $\boldsymbol{\Sigma}$ has diagonal elements σ_j^2 and off-diagonal elements $\rho\sigma_j\sigma_k$; thus $\boldsymbol{\Sigma} = \mathbf{s}\{(1 - \rho)\mathbf{I}_p + \rho\mathbf{J}_p\}\mathbf{s}$ where $\mathbf{s} = \mathrm{diag}(\sigma_1,\ldots,\sigma_p)$. Each pair of components of the error vector $\mathbf{e}_i$ has correlation ρ. The kind of situation for which this structure may be appropriate is where the measurements are all made at about the same time, as in the often used (but seldom explained) 'split-plots' set-up, rather than in some strict temporal sequence; the latter case is dealt with below. For measurements of the same type made in the same way it is usual to take $\sigma_j^2 = \sigma^2$ for all j, i.e. to assume variance homogeneity.

Note that $\boldsymbol{\Sigma}^{-1} = (1 - \rho)^{-1}\mathbf{s}^{-1}\{\mathbf{I}_p - (\rho/a)\mathbf{J}_p\}\mathbf{s}^{-1}$ and $\det\boldsymbol{\Sigma} = (\det\mathbf{s})^2(1 - \rho)^{p-1}a$, where $a = 1 + (p - 1)\rho$. These forms are useful for computation of the likelihood function.

5.4.2 Autoregressive errors

When the measurements on an individual have been made in sequence over time the errors may be correlated. Thus, if it were possible to run many independent repetitions of the process on the same individual i, one could discern some consistent pattern linking the components e_j of the error vectors in single runs. A widely used time series model is the autoregressive process

$$e_j = \rho e_{j-1} + u_j \qquad \text{for } j > 2 \qquad (5.4.1)$$

where ρ is the regression parameter and the u_j are innovation errors usually assumed to be $N(0, \sigma^2)$ variates each independent of the past.

Repeated application of (5.4.1) for $e_{j-1}, e_{j-2}, \ldots$, yields

$$e_j = u_j + \rho u_{j-1} + \rho^2 u_{j-2} + \cdots + \rho^{k-1} u_{j-k+1} + \rho^k e_{j-k} \qquad \text{for } k > 0.$$
(5.4.2)

There are two associated submodels. In the first the process (5.4.1) is assumed to have been running long before observation started, i.e. since $j = -\infty$, and to be 'stable'. This means that $|\rho| < 1$ and the e_j all have the same distribution, which is therefore normal with mean zero and a common variance. From (5.4.2)

$$V(e_j) = \sigma^2(1 + \rho^2 + \rho^4 + \cdots + \rho^{2(k-1)}) + \rho^{2k} V(e_{j-k})$$

and so this common variance is $V(e_j) = \sigma^2/(1 - \rho^2)$. Also from (5.4.2), $E(e_j e_{j-k}) = \rho^k E(e_{j-k}) = \sigma^2 \rho^k/(1 - \rho^2)$, and it follows that $(\mathbf{\Sigma})_{jk} = \sigma^2 \rho^{|j-k|}/(1 - \rho^2)$. Then $\mathbf{\Sigma}^{-1}$ is a tridiagonal matrix with diagonal $\sigma^2(1, 1 + \rho^2, 1 + \rho^2, \ldots, 1 + \rho^2, 1)$ and subdiagonal $-\sigma^{-2}(\rho, \ldots, \rho)$; also, $\det \mathbf{\Sigma} = \sigma^{2p}/(1 - \rho^2)$.

In the second submodel the process (5.4.1) is deemed to have started up with the first observation, i.e. at $j = 1$, with $u_0 = 0$ and $e_1 = u_1$. From (5.4.2) one can obtain the covariance structure now as $(\mathbf{\Sigma})_{jk} = \sigma^2 \rho^{j-k}(1 - \rho^{2k})/(1 - \rho^2)$ for $j \geq k$. In contrast to the stable submodel, the e_j now have unequal variances and there is no restriction to $|\rho| < 1$. Here $\mathbf{\Sigma}^{-1}$ is tridiagonal with diagonal $\sigma^{-2}(1 + \rho^2, 1 + \rho^2, \ldots, 1 + \rho^2, 1)$ and subdiagonal $-\sigma^{-2}(\rho, \ldots, \rho)$, and $\det \mathbf{\Sigma} = \sigma^{2p}$. Thus the inverse matrices $\mathbf{\Sigma}^{-1}$ for the two submodels differ only in the $(1, 1)$ entry.

5.4.3   Markov correlation structure

The stable autoregressive submodel in section 5.4.2, which is a homogeneous Markov chain, has correlations $\rho^{|j-k|}$. Thus the correlation between components e_j decreases geometrically with their separation in time. When the data are unequally spaced over time, the measurement occasions being at times t_j, say, the geometrically decreasing correlation form can be sensibly incorporated by taking $(\mathbf{\Sigma})_{jk} = \sigma^2 \rho^{|t_j - t_k|}$. This form is invariant under linear transformation of the time scale. For instance, if one changes from weeks to days and moves the origin to day 3, then $t \to 7t - 3$ but $\rho \to \rho^{1/7}$ compensates. By the same token, however, the magnitude of ρ just depends on the time scale used. Usually $0 \leq \rho < 1$, but in some cases, e.g. when the $|t_j - t_k|$ are integers, one can allow

$-1 < \rho < 1$. Here Σ^{-1} has tridiagonal form with diagonal $\sigma^{-2}(f_1,$ $f_1 + f_2 - 1, \ldots, f_{p-2} + f_{p-1} - 1, \quad f_{p-1})$ and subdiagonal $-\sigma^{-2}$ $(f_1 \rho^{d_1}, \ldots, f_{p-1} \rho^{d_{p-1}})$ where $d_j = t_{j+1} - t_j$ and $f_j = (1 - \rho^{2d_j})^{-1}$; also det $\Sigma = \sigma^{2p}(f_1 f_2 \cdots f_{p-1})^{-1}$. For the record, setting $z = (z_1, \ldots, z_p)^T$, one can derive $z^T \Sigma^{-1} z = f_1(z_1 - \rho^{d_1} z_2)^2 + f_2(z_2 - \rho^{d_2} z_3)^2 + \cdots + f_{\rho - 1} z_\rho^2$ to verify that Σ^{-1} (and therefore Σ) is positive definite.

5.5 Further reading

Section 5.1 is based on Rao (1959) who gives a more sophisticated analysis. Effectively, he has $n = 1$ and assumes that an independent estimate of Σ is available.

Grizzle and Allen (1969), following on from Potthoff and Roy (1964) and Khatri (1966), give an account of a particular growth curve framework. It corresponds to the model of section 5.2 with $p_i = p$ and a certain special form for X_i reflecting a between- and within-groups design. Azzalini (1987) generalizes the analysis to the case of structured Σ. In particular, he examines autoregressive errors and random regression coefficients (see Chapter 6). Verbyla and Venables (1988) extend the Potthoff–Roy model to cover cases where treatment groups may have different profile forms. Srivastava and McDonald (1974) generalize Potthof and Roy's analysis in a different direction, to cope with a certain pattern of missing values, i.e. where some data vectors may be cut short but not contain gaps. Crépeau et al. (1985) analyse two examples of this type of incomplete data, the designs being one-way layouts in each case. Patel and Khatri (1981) also consider such data, which they call 'monotone samples', concentrating on a one-way layout with a covariate and a Markov-structured covariance matrix Σ.

For the model of section 5.3 Jennrich and Schluchter (1986) compare various computational schemes and fit Potthoff and Roy's (1964) data with various Σ structures. Mukherjee (1976) gives a useful review of the field with many references. Ware's (1985) review concentrates on three particular Σ-structures: multivariate (section 5.1), random effects (section 6.1) and autoregressive (section 5.4.2).

Particular covariance structures have been studied by many authors. One important class will be the subject of Chapter 6. Rao (1967) considers least-squares estimators with this class together with unstructured and autoregressive forms. Berenblut and Webb (1973) apply the nonstationary autoregressive model of section 5.4.2. Beach and MacKinnon (1978) look at maximum likelihood estimation in the autoregressive case. Wilson, Hebel and Sherwin (1981) consider, among other things, m.l.e. for the model $\mathbf{y}_i = \mu \mathbf{1}_p + \mathbf{e}_i$ with $\mathbf{\Sigma}_i = \sigma_u^2 \mathbf{J}_p + \mathbf{V}$, $\mathbf{V}$ having the autoregressive form of section 5.4.2. Azzalini (1984) adds covariates to this scheme. Kenward (1985) uses the same covariance structure to investigate the efficacy of using higher-order orthogonal polynomial coefficients as covariates in growth curve analysis. Rochon and Helms (1989) cover m.l.e. for a linear model with an ARMA covariance structure for repeated measures with missing values. Srivastava (1984) and Srivastava and Keen (1988) consider inference about the parameters of the model $\mathbf{y}_i \sim N(\mathbf{\mu}_i, \mathbf{\Sigma}_i)$ with $\mathbf{\mu}_i^{\mathrm{T}} = (\mu_m, \mu_c \mathbf{1}_p^{\mathrm{T}})$, representing a mother and a number of children, and $\mathbf{\Sigma}_i = \begin{pmatrix} \sigma_m^2 & \sigma_{mc} \mathbf{1}_p^{\mathrm{T}} \\ \sigma_{mc} \mathbf{1}_p & \sigma_c^2 \mathbf{\Phi}_i \end{pmatrix}$, $\mathbf{\Phi}_i = (1 - \rho_{ss}) \mathbf{I}_p + \rho_{ss} \mathbf{J}_p$ (an equicorrelated structure).

An important class of covariance forms is that of 'linear structure' introduced by Anderson (1969, 1970, 1973). Here $\mathbf{\Sigma}$ or $\mathbf{\Sigma}^{-1}$ is expressible as a sum of terms $\phi_j \mathbf{G}_j$ where the matrices $\mathbf{G}_j$ are known but the scalars ϕ_j are not. Rogers and Young (1977) give some further results for this set-up, and Andrade and Helms (1984) suggest computation based on the EM algorithm.

Anderson's (1978) paper, despite its title, is really about certain multivariate time series.

CHAPTER 6

Two-stage linear models

6.1 Random regression coefficients

We come now to what is perhaps the most satisfying, and sometimes the most satisfactory, approach to repeated measurements, at least from the regression modelling point of view. It is essentially a generalization of the basic Anova model (3.1.1) of Chapter 3. The fixed effects means $\mu_{ij} = E(y_{ij})$ are extended to a general regression model $\boldsymbol{\mu}_i = \mathbf{X}_i\boldsymbol{\beta}$ as in Chapter 5, and the random effects α_{ij} similarly to $\mathbf{Z}_i\mathbf{b}_i$. The resulting covariance structure is more general than that of compound symmetry in Chapter 3, but more naturally restricted than the completely undisciplined form of sections 5.1 and 5.2. The ideas will be introduced via an idealized example.

Consider the straight line regression $y_{ij} = \alpha + \beta x_j + e_{ij}$, where y_{ij} is the jth measurement on the ith individual. Suppose that there is some variation in the slope between individuals, so β should be replaced by $\beta + b_i$ where β is the population mean slope and b_i is the individual departure from this mean slope. Thus $E(b_i) = 0$, the expectation being taken over the population of individuals. Also suppose for a moment that α does not vary over individuals. Assuming that the errors e_{ij} are uncorrelated, both with each other and with the b_i, we have

$$E(y_{ij}) = \alpha + \beta x_j, \ V(y_{ij}) = \sigma_b^2 x_j^2 + \sigma^2, C(y_{ij}, y_{ij'}) = \sigma_b^2 x_j x_{j'} \text{ for } j \neq j'$$

where $\sigma_b^2 = V(b_i)$ and $\sigma^2 = V(e_{ij})$. Note that this rather natural, not overly elaborate or technical, concept of variation in the slopes has conjured up both heterogeneity of variance ($V(y)$ increases with x) and correlated observations (y_{ij} with $y_{ij'}$) at a stroke.

Suppose now that the intercept α also varies over individuals, so we replace it by $\alpha + a_i$ where $E(a_i) = 0$ and $V(a_i) = \sigma_a^2$ (over the population of individuals); also let $C(a_i, b_i) = \sigma_{ab}$ and assume that

$C(a_i, e_{ij}) = 0$. The consequences now for y_{ij} are that

$$E(y_{ij}) = \alpha + \beta x_j, \; V(y_{ij}) = (\sigma_a^2 + 2x_j\sigma_{ab} + x_j^2\sigma_b^2) + \sigma^2,$$

$$C(y_{ij}, y_{ij'}) = \sigma_a^2 + (x_j + x_{j'})\sigma_{ab} + x_j x_{j'}\sigma_b^2 \text{ for } j \neq j'.$$

These formulae are much tidier in vector and matrix notation. Take $\mathbf{y}_i = (y_{i1}, \ldots, y_{ip})^{\mathrm{T}}$, $\boldsymbol{\beta} = (\alpha, \beta)^{\mathrm{T}}$, $\mathbf{b}_i = (a_i, b_i)^{\mathrm{T}}$, $\mathbf{e}_i = (e_{i1}, \ldots, e_{ip})^{\mathrm{T}}$ and $\mathbf{X}(p \times 2)$ with jth row $(1, x_j)$. Then we have $\mathbf{y}_i = \mathbf{X}(\boldsymbol{\beta} + \mathbf{b}_i) + \mathbf{e}_i$, $\boldsymbol{\mu}_i = E(\mathbf{y}_i) = \mathbf{X}\boldsymbol{\beta}$ and $\boldsymbol{\Sigma}_i = V(\mathbf{y}_i) = \mathbf{X}\mathbf{B}\mathbf{X}^{\mathrm{T}} + \mathbf{E}$ where

$$\mathbf{B} = V(\mathbf{b}_i) = \begin{pmatrix} \sigma_a^2 & \sigma_{ab} \\ \sigma_{ab} & \sigma_b^2 \end{pmatrix}$$

and $\mathbf{E} = V(\mathbf{e}_i) = \sigma^2 \mathbf{I}_p$. Note in particular the structured form for $\boldsymbol{\Sigma}_i$.

The model will now be broadened out from the example by first extending $\boldsymbol{\beta}$ to a $q \times 1$ vector of regression coefficients, and replacing p by p_i to allow differing numbers of measurements on different individuals. Hence it becomes $\mathbf{y}_i = \mathbf{X}_i\boldsymbol{\beta} + \mathbf{X}_i\mathbf{b}_i + \mathbf{e}_i$, with $\mathbf{X}_i$ now $p_i \times q$. Further, $\mathbf{X}_i\mathbf{b}_i$ will be changed to $\mathbf{Z}_i\mathbf{b}_i$, where $\mathbf{Z}_i(p_i \times r)$ may differ from $\mathbf{X}_i$, and $\mathbf{b}_i(r \times 1)$ has mean $\mathbf{0}$ and covariance matrix $\mathbf{B}$ over the population of individuals. The introduction of $\mathbf{Z}_i$ distinct from $\mathbf{X}_i$, though not strictly necessary, allows neater separation of the fixed ($\mathbf{X}_i\boldsymbol{\beta}$) from the random ($\mathbf{Z}_i\mathbf{b}_i$) effects. Comparing this very general regression model with the basic Anova form (3.1.1) the progression from $(\mu_{ij}, \alpha_{ij}, e_{ij})$ to $(\mathbf{X}_i\boldsymbol{\beta}, \mathbf{Z}_i\mathbf{b}_i, \mathbf{e}_i)$ is clear. To summarize, the model is

$$\mathbf{y}_i = \mathbf{X}_i\boldsymbol{\beta} + \mathbf{Z}_i\mathbf{b}_i + \mathbf{e}_i \tag{6.1.1a}$$

together with the following specifications on the random terms:

$$E(\mathbf{e}_i) = \mathbf{0}, \quad V(\mathbf{e}_i) = \mathbf{E}_i, \tag{6.1.1b}$$

$$E(\mathbf{b}_i) = \mathbf{0}, \quad V(\mathbf{b}_i) = \mathbf{B}, \quad C(\mathbf{b}_i, \mathbf{e}_i) = 0 \tag{6.1.1c}$$

It is understood implicitly that random quantities from different individuals are uncorrelated, i.e. $C(\mathbf{b}_i, \mathbf{b}_{i'}) = \mathbf{0}, C(\mathbf{e}_i, \mathbf{e}_{i'}) = \mathbf{0}$ and $C(\mathbf{b}_i, \mathbf{e}_{i'}) = \mathbf{0}$ for $i \neq i'$. In some situations this consideration can serve actually to define 'individuals', in the sense of deciding how the measurements are to be grouped into uncorrelated vectors. The framework (6.1.1) is often referred to as a **two-stage model**, the first being within-individuals (6.1.1a and 6.1.1b) and the second being across-individuals (6.1.1c). In the first stage $\mathbf{b}_i$ is regarded as a fixed parameter for the individual whose measurements are being described, and then in the second stage one describes the whole population of individuals and the variation of $\mathbf{b}_i$ over it. The consequence of (6.1.1) for $\mathbf{y}_i$ is

$$\boldsymbol{\mu}_i = E(\mathbf{y}_i) = \mathbf{X}_i\boldsymbol{\beta}, \qquad \boldsymbol{\Sigma}_i = V(\mathbf{y}_i) = \mathbf{Z}_i\mathbf{B}\mathbf{Z}_i^T + \mathbf{E}_i \qquad (6.1.2)$$

and the implicit understanding is that $C(\mathbf{y}_i, \mathbf{y}_{i'}) = \mathbf{0}$.

Only the means, variances and covariances are involved in (6.1.1) and (6.1.2). It is usual for statistical analysis further to assume normal distributions for the $\mathbf{b}_i$ and $\mathbf{e}_i$, and consequently for the $\mathbf{y}_i$. Thus (6.1.2) is strengthened to the statement that:

the $\mathbf{y}_i$ are independently $N_p(\mathbf{X}_i\boldsymbol{\beta}, \boldsymbol{\Sigma}_i)$ with $\boldsymbol{\Sigma}_i = \mathbf{Z}_i\mathbf{B}\mathbf{Z}_i^T + \mathbf{E}_i$. (6.1.3)

Thus (6.1.3) is covered by the general regression model of section 5.3 with $\boldsymbol{\phi}$ containing all the parameters occurring in $\mathbf{B}$ and the $\mathbf{E}_i$.

In applications $\mathbf{E}_i$ is taken to have some special simple form depending on a small number of parameters, otherwise (6.1.3) would impose no constraint on $\boldsymbol{\Sigma}_i$. For example, $\mathbf{E}_i = \sigma^2\mathbf{I}_p$ is a common choice. The $r \times r$ matrix $\mathbf{B}$ may also be economically parametrized, but in any case will involve at most $r(r + 1)/2$ unknowns. Since $\boldsymbol{\Sigma}_i$ is $p_i \times p_i$, with $p_i(p_i + 1)/2$ distinct elements, and p_i is usually substantially greater than r, one can see that the constraint on $\boldsymbol{\Sigma}_i$ in (6.1.3) represents a considerable saving in parameters compared with the unconstrained multivariate approach of section 5.2. Note that $\mathbf{B}$ has not been assumed to be nonsingular. Particular forms for $\mathbf{E}_i$ and $\mathbf{B}$, including singular $\mathbf{B}$ and $\mathbf{E}_i = \sigma^2\mathbf{I}_{p_i}$, will be discussed in section 6.3. It is common to assume diagonal form for $\mathbf{E}_i$ and to refer to this as the **conditional independence model.**

6.2 Estimation and testing

In practice, models of type (6.1.1) will often be constructed as a result of investigating individual profiles, as in section 2.3. The formal fitting and testing of an overall model, as described now, is just an application of the general methods of section 5.3.

The generalized least-squares estimator for $\boldsymbol{\beta}$ is given by

$$\hat{\boldsymbol{\beta}} = \left(\sum_{i=1}^{n} \mathbf{X}_i^T\boldsymbol{\Sigma}_i^{-1}\mathbf{X}_i \right)^{-1} \left(\sum_{i=1}^{n} \mathbf{X}_i^T\boldsymbol{\Sigma}_i^{-1}\mathbf{y}_i \right) \qquad (6.2.1)$$

as in (5.2.2), but with the particular form (6.1.2) for $\boldsymbol{\Sigma}_i$ here. The covariance matrix of $\hat{\boldsymbol{\beta}}$ is $\mathbf{V}_\beta = (\sum_{i=1}^{n} \mathbf{X}_i^T\boldsymbol{\Sigma}_i^{-1}\mathbf{X}_i)^{-1}$, again as in section 5.2. Maximum likelihood estimates may be found as described in section 5.3.

When $\mathbf{B}$ is nonsingular the formula (Rao, 1965, p. 29)

$$\mathbf{\Sigma}^{-1} = \mathbf{E}^{-1} - \mathbf{E}^{-1}\mathbf{Z}(\mathbf{Z}^T\mathbf{E}^{-1}\mathbf{Z} + \mathbf{B}^{-1})^{-1}\mathbf{Z}^T\mathbf{E}^{-1} \qquad (6.2.2)$$

may be usefully employed in some circumstances. If p_i is large direct inversion of $\mathbf{\Sigma}_i$ for (6.2.1) is generally costly in computation. On the other hand, (6.2.2) requires $\mathbf{E}^{-1}$ and $\mathbf{B}^{-1}$. When r is moderate $\mathbf{B}^{-1}$ will be cheaply computed, and likewise $\mathbf{E}_i^{-1}$ in the common situation where $\mathbf{E}_i$ has simple form such as $\sigma^2\mathbf{I}_p$. A similarly useful formula for $\det\mathbf{\Sigma}_i$ is

$$\det\mathbf{\Sigma} = (\det\mathbf{E})(\det\mathbf{B})\det(\mathbf{B}^{-1} + \mathbf{Z}^T\mathbf{E}^{-1}\mathbf{Z}). \qquad (6.2.3)$$

Unfortunately, it is not uncommon for some parameters to go out of bounds during the iterative search for a maximum. For instance, $\mathbf{B}$ may become non positive-semi-definite at some point. One way of avoiding such problems is to parametrize the various quantities so that any constraints are satisfied automatically. Thus, $\mathbf{B}$ can always be expressed in the form $\mathbf{AA}^T$, for some appropriately chosen $\mathbf{A}$, and this will often provide a workable parametrization for $\mathbf{B}$.

As in section 5.2 standard theory yields $\hat{\mathbf{\beta}} \xrightarrow{D} N_q(\mathbf{\beta}, \mathbf{V}_\beta)$ as $n \to \infty$, and $\hat{\mathbf{V}}_\beta = (\sum_{i=1}^{n}\mathbf{X}_i^T\hat{\mathbf{\Sigma}}_i^{-1}\mathbf{X}_i)^{-1} \xrightarrow{P} \mathbf{V}_\beta$, where $\hat{\mathbf{\Sigma}}_i = \mathbf{Z}_i\hat{\mathbf{B}}\mathbf{Z}_i + \hat{\mathbf{E}}_i$. Hence, tests and confidence intervals can be constructed for $\hat{\mathbf{\beta}}$ based on the large-sample approximation $N_q(\mathbf{\beta}, \hat{\mathbf{V}}_\beta)$ to the distribution of $\hat{\mathbf{\beta}}$. Residuals may be defined and examined as described in section 5.3.

In some situations estimates of the individual parameters $\mathbf{b}_i$ are required as well as of $\mathbf{\beta}$. For example, one might wish to predict for an individual, or pick out one with some extreme characteristic that is governed by $\mathbf{b}_i$. This can be done within an extended likelihood framework as follows.

Consider the conditional distribution of $\mathbf{y}_i$ given $\mathbf{b}_i$ in (6.1.1a), i.e. with variation of $\mathbf{e}_i$ but sticking with the same individual so that $\mathbf{b}_i$ is fixed like $\mathbf{\beta}$. This conditional distribution is $N_{p_i}(\mathbf{X}_i\mathbf{\beta} + \mathbf{Z}_i\mathbf{b}_i, \mathbf{E}_i)$ with density

$$f(\mathbf{y}_i|\mathbf{b}_i) = [\det(2\pi\mathbf{E}_i)]^{-1/2}\exp(-Q_i/2) \qquad (6.2.4)$$

where $Q_i = (\mathbf{y}_i - \mathbf{X}_i\mathbf{\beta} - \mathbf{Z}_i\mathbf{b}_i)^T\mathbf{E}_i^{-1}(\mathbf{y}_i - \mathbf{X}_i\mathbf{\beta} - \mathbf{Z}_i\mathbf{b}_i)$. Now the marginal distribution of $\mathbf{b}_i$ is $N_r(\mathbf{0}, \mathbf{B})$ with density

$$f(\mathbf{b}_i) = [\det(2\pi\mathbf{B})]^{-1/2}\exp(-\mathbf{b}_i^T\mathbf{B}^{-1}\mathbf{b}_i), \qquad (6.2.5)$$

assuming $\mathbf{B}$ to be nonsingular to avoid unimportant complications.

Hence the joint density $f(\mathbf{y}_i, \mathbf{b}_i)$ of $\mathbf{y}_i$ and $\mathbf{b}_i$ is obtained as the product of (6.2.4) and (6.2.5). The marginal density of $\mathbf{y}_i$, on which the likelihood function was based above, could be obtained by integrating $f(\mathbf{y}_i, \mathbf{b}_i)$ over $\mathbf{b}_i$. The resulting $N_{p_i}(\mathbf{X}_i\boldsymbol{\beta}_i, \boldsymbol{\Sigma}_i)$ distribution for $\mathbf{y}_i$, when compared with (6.1.3), verifies (6.2.2) as noted by Lindley and Smith (1972, section 2).

The joint likelihood for $(\mathbf{y}_i, \mathbf{b}_i)(i = 1, \ldots, n)$ is

$$L_n = \log \prod_{i=1}^{n} f(\mathbf{y}_i, \mathbf{b}_i) = -\tfrac{1}{2}(p_+ + nr)\log(2\pi)$$

$$-\tfrac{1}{2}\{\log(\det \mathbf{E}_i) + \log(\det \mathbf{B})\} - \tfrac{1}{2}\sum_{i=1}^{n}(Q_i + \mathbf{b}_i^T \mathbf{B}^{-1}\mathbf{b}_i).$$

$$(6.2.6)$$

Maximization of L_n over $\boldsymbol{\beta}$ and $\mathbf{b}_1, \ldots, \mathbf{b}_n$ leads to the estimates

$$\hat{\mathbf{b}}_i = (\mathbf{B}^{-1} + \mathbf{Z}_i^T \mathbf{E}_i^{-1}\mathbf{Z}_i)^{-1}\mathbf{Z}_i^T \mathbf{E}_i^{-1}(\mathbf{y}_i - \mathbf{X}_i\hat{\boldsymbol{\beta}})$$

$$= \mathbf{B}\mathbf{Z}_i^T \boldsymbol{\Sigma}_i^{-1}(\mathbf{y}_i - \mathbf{X}_i\hat{\boldsymbol{\beta}}) \qquad (6.2.7)$$

for the $\mathbf{b}_i$, and $\hat{\boldsymbol{\beta}}$ as in (6.2.1) for $\boldsymbol{\beta}$. For known $\boldsymbol{\phi}$ the variance of $\hat{\boldsymbol{\beta}}$ is $\mathbf{V}_\beta$ as given above, that of $\hat{\mathbf{b}}_i$ is

$$V(\hat{\mathbf{b}}_i) = \mathbf{B}\mathbf{Z}_i^T(\boldsymbol{\Sigma}_i^{-1} - \boldsymbol{\Sigma}_i^{-1}\mathbf{X}_i\mathbf{V}_\beta\mathbf{X}_i^T\boldsymbol{\Sigma}_i^{-1})\mathbf{Z}_i\mathbf{B} = \mathbf{V}_i$$

say, and that of $(\hat{\mathbf{b}}_i - \mathbf{b}_i)$ comes down to $\mathbf{B} - \mathbf{V}_i$. To assess how good an estimator $\hat{\mathbf{b}}_i$ is of $\mathbf{b}_i$ the appropriate variance is $V(\hat{\mathbf{b}}_i - \mathbf{b}_i)$, not $V(\hat{\mathbf{b}}_i)$. To qualify a prediction $\mathbf{a}_1^T\hat{\boldsymbol{\beta}} + \mathbf{a}_2^T\hat{\mathbf{b}}_i$ of a given linear function $\mathbf{a}_1^T\boldsymbol{\beta} + \mathbf{a}_2^T\mathbf{b}_i$ the appropriate variance is

$$V[(\mathbf{a}_1^T\hat{\boldsymbol{\beta}} + \mathbf{a}_2^T\hat{\mathbf{b}}_i) - (\mathbf{a}_1^T\boldsymbol{\beta} + \mathbf{a}_2^T\mathbf{b}_i)]$$

$$= \mathbf{a}_1^T\mathbf{V}_\beta\mathbf{a}_1 + 2\mathbf{a}_1^T C(\hat{\boldsymbol{\beta}}, \hat{\mathbf{b}}_i - \mathbf{b}_i)\mathbf{a}_2 + \mathbf{a}_2^T(\mathbf{B} - \mathbf{V}_i)\mathbf{a}_2$$

where

$$C(\hat{\boldsymbol{\beta}}, \hat{\mathbf{b}}_i - \mathbf{b}_i) = -\mathbf{V}_\beta\mathbf{X}_i^T\boldsymbol{\Sigma}_i^{-1}\mathbf{Z}_i\mathbf{B}.$$

6.3 Particular aspects

6.3.1 An Anova-type covariance structure

Let us focus on the $\mathbf{Z}_i\mathbf{B}\mathbf{Z}_i^T$ part of $\boldsymbol{\Sigma}_i$ in (6.1.1), the usual form for $\mathbf{E}_i$ in Anova being $\sigma^2\mathbf{I}_p$. The examples of Chapter 3 will be used for illustration.

The model of Example 3.1, in vector form is

$$\mathbf{y}_i = \boldsymbol{\mu}^{(g)} + \alpha_i^{\mathrm{I}}\mathbf{1}_p + \boldsymbol{\alpha}_i^{\mathrm{IO}} + \mathbf{e}_i$$

where $\boldsymbol{\mu}^{(g)}$ represents the mean weight profile for group g and $\boldsymbol{\alpha}_i^{\mathrm{IO}} = (\alpha_{i1}^{\mathrm{IO}}, \ldots, \alpha_{ip}^{\mathrm{IO}})^{\mathrm{T}}$ contains the random *individual $\times$ occasion* interactions. The random part $\alpha_i^{\mathrm{I}}\mathbf{1}_p + \boldsymbol{\alpha}_i^{\mathrm{IO}}$ can be represented as $\mathbf{Z}_i\mathbf{b}_i$ by taking $r = p + 1$, $\mathbf{b}_i = \begin{pmatrix} \alpha_i^{\mathrm{I}} \\ \boldsymbol{\alpha}_i^{\mathrm{IO}} \end{pmatrix}$ and $\mathbf{Z}_i = (\mathbf{1}_p, \mathbf{I}_p)$. Then $\mathbf{B} = \begin{pmatrix} \sigma_{\mathrm{I}}^2 & \mathbf{0}_p^{\mathrm{T}} \\ \mathbf{0}_p & \sigma_{\mathrm{IO}}^2\mathbf{K}_p \end{pmatrix}$
where $\mathbf{0}_p = (0, \ldots, 0)^{\mathrm{T}}$ and $\mathbf{K}_p = (p\mathbf{I}_p - \mathbf{J}_p)/(p-1)$ has (j, k)th element 1 for $j = k$ and $-1/(p-1)$ for $j \neq k$. The off-diagonal elements of $\mathbf{K}_p$, the correlation matrix of $\boldsymbol{\alpha}_i^{\mathrm{IO}}$, are determined by the constraint $\alpha_{i+}^{\mathrm{IO}} = 0$. This makes $\mathbf{K}_p$, and hence $\mathbf{B}$, singular. To avoid this one could take instead $r = p$, $\mathbf{b}_i^{\mathrm{T}} = (\alpha_i^{\mathrm{I}}, \alpha_{i1}^{\mathrm{IO}}, \ldots, \alpha_{i,p-1}^{\mathrm{IO}})$, and modify $\mathbf{Z}_i$ by deleting the last column and changing its last row to $(1, -1, -1, \ldots, -1)$. In either case $\mathbf{Z}_i\mathbf{b}_i$, and hence $\mathbf{Z}_i\mathbf{B}\mathbf{Z}_i^{\mathrm{T}}$, the quantities actually appearing in the $\mathbf{y}_i$-model, are the same. Explicitly, $\mathbf{Z}_i\mathbf{B}\mathbf{Z}_i^{\mathrm{T}} = \sigma_{\mathrm{I}}^2\mathbf{J}_p + \sigma_{\mathrm{IO}}^2\mathbf{K}_p$.

In Example 3.2, where each subject is observed over an $a \times b$ factorial design, the random effects comprise $\alpha_i^{\mathrm{I}} + \alpha_{ij}^{\mathrm{IA}} + \alpha_{ik}^{\mathrm{IB}} + \alpha_{ijk}^{\mathrm{IAB}}$. Write

$$\mathbf{b}_i = (\alpha_i^{\mathrm{I}} | \alpha_{i1}^{\mathrm{IA}}, \ldots, \alpha_{ia}^{\mathrm{IA}} | \alpha_{i1}^{\mathrm{IB}}, \ldots, \alpha_{ib}^{\mathrm{IB}} | \alpha_{i11}^{\mathrm{IAB}}, \alpha_{i12}^{\mathrm{IAB}}, \ldots, \alpha_{iab}^{\mathrm{IAB}})$$

with $r = 1 + a + b + ab$. Then $\mathbf{Z}_i$ must consist of 1's and 0's arranged with devilish ingenuity to pick out the appropriate α's from $\mathbf{b}_i$. Also $\mathbf{B}$ will be block-diagonal with blocks (σ_{I}^2), $\sigma_{\mathrm{IA}}^2\mathbf{K}_a$, $\sigma_{\mathrm{IB}}^2\mathbf{K}_b$ and $\sigma_{\mathrm{IAB}}^2\mathbf{L}$ where $\mathbf{K}_a$ and $\mathbf{K}_b$ are like $\mathbf{K}_p$ in the previous paragraph; $\mathbf{L}(ab \times ab)$ comprises an $a \times a$ array of $b \times b$ blocks, those on the diagonal being $\mathbf{K}_b$ and those off it being $-\mathbf{K}_b/(a-1)$. The form of $\mathbf{L}$ follows from the discussion in sections 3.5 and 3.6.

These examples illustrate the advantage of uncoupling $\mathbf{Z}_i$ from $\mathbf{X}_i$ in the model (6.1.1). Thus, $\mathbf{X}_i$ can be used to represent the various fixed effects, including 0/1 experimental design indicators and continuous x-covariates, while $\mathbf{Z}_i$ is used independently to construct a desired covariance pattern.

6.3.2. Growth curves

Much of the classic work on two-stage models concerns the fitting of polynomial curves to animal growth measurements over time. Here $\mathbf{X}$ has jth row $(1, x_j, x_j^2, \ldots, x_j^{q-1})$, or the 'orthogonal polynomial' version of this. Also, in that literature, $\mathbf{X}_i$ has generally been equal to $\mathbf{Z}_i$ or $\mathbf{Z}_i\mathbf{A}_i$ for some $\mathbf{A}_i$. This arises from $\mathbf{y}_i = \mathbf{Z}_i\boldsymbol{\beta}_i + \mathbf{e}_i$ at the first stage, and then $\boldsymbol{\beta}_i$ having distribution $\mathrm{N}_q(\mathbf{A}_i\boldsymbol{\beta}, \mathbf{B})$ at the second. Thus $E(\mathbf{y}_i) = \mathbf{X}_i\boldsymbol{\beta}$, with $\mathbf{X}_i = \mathbf{Z}_i\mathbf{A}_i$, and $V(\mathbf{y}_i) = \mathbf{Z}_i\mathbf{B}\mathbf{Z}_i^{\mathrm{T}} + \mathbf{E}_i$. The introductory example in section 6.1 was of this type with $\boldsymbol{\beta}_i = (\alpha + a_i, \beta + b_i)$, $\mathbf{Z}_i$ having jth row $(1, x_j)$ and $\mathbf{A}_i = \mathbf{I}_2$.

For polynomial growth curves many authors have noted the possibility that individual curves may be of higher degree than the population mean curve. This occurs when the latter components of $\boldsymbol{\beta}_i$ have zero means.

Laird and Ware (1982) draw the following distinction between growth curves and repeated measures. In the former the natural development or ageing process of the individual tends to be monitored without intervention, and comparisons are made between different groups. In the latter, individual characteristics tend to be constant over the period of study while the individual undergoes different treatments or conditions. Of course, the distinction is not rigid, and Example 2.4 here provides a ready exception.

When $\mathbf{X}_i = \mathbf{Z}_i$ there is an interesting relationship between $\hat{\boldsymbol{\beta}}$, the least-squares estimator in (6.2.1), and the set of individually estimated regression coefficients $\hat{\boldsymbol{\beta}}_i$; the latter, defined as $\hat{\boldsymbol{\beta}}_i = (\mathbf{X}_i^{\mathrm{T}}\mathbf{E}_i^{-1}\mathbf{X}_i)^{-1}(\mathbf{X}_i^{\mathrm{T}}\mathbf{E}_i^{-1}\mathbf{y}_i)$, is the least-squares estimator from the data on the ith individual alone. The relationship is

$$\hat{\boldsymbol{\beta}} = \mathbf{V}_\beta^{-1} \sum_{i=1}^n \mathbf{X}_i^{\mathrm{T}}\boldsymbol{\Sigma}_i^{-1}\mathbf{X}_i\hat{\boldsymbol{\beta}}_i,$$

which shows $\hat{\boldsymbol{\beta}}$ as a weighted average of the $\hat{\boldsymbol{\beta}}_i$ with weighting matrices $\mathbf{X}_i^{\mathrm{T}}\boldsymbol{\Sigma}_i^{-1}\mathbf{X}_i$. It may be derived from (6.2.1) after using (6.2.2) to verify that $\mathbf{X}_i^{\mathrm{T}}\boldsymbol{\Sigma}_i^{-1}\mathbf{X}_i\hat{\boldsymbol{\beta}}_i = \mathbf{X}_i^{\mathrm{T}}\boldsymbol{\Sigma}_i^{-1}\mathbf{y}_i$.

6.3.3 Multilevel models

Goldstein (1979, 1986a, b, 1987) and others have been applying multilevel models to education data over the past few years. Goldstein's (1987) book gives an extended account, with thorough discussion of all aspects, and the present subsection is just intended to show how such models fit into the general framework (6.1.1).

Recall the fundamental model (3.1.1):

$$y_{ij} = \mu_{ij} + \alpha_{ij} + e_{ij}$$

and suppose that y_{ij} is the score of child j in class i in a school examination. The fixed effects μ_{ij} may take a regression form $\beta_0 + \beta_1 x_{ij} + \beta_2 x_i$, where x_{ij} is an explanatory variable associated with the child (e.g. a previous test score) and x_i is one associated with the class (e.g. its size); of course, several such covariates may be present, in which case β_1, x_{ij}, β_2 and x_i become vectors. For the random effects the simplest form $\alpha_{ij} = \alpha_i^C$, representing a random individual class effect, may obtain. Goldstein (1987, section 2.1) gives essentially this example, and calls it a **two-level** model because the random terms e_{ij} (at level 1, the child) and α_i^C (at level 2, the class), are at two levels. In terms of the present framework the model is covered by (6.1.1) with y_{ij}, μ_{ij} and e_{ij} forming the components respectively of the vectors $\mathbf{y}_i$, $\mathbf{X}_i\boldsymbol{\beta}_i$ and $\mathbf{e}_i$, and with $\mathbf{Z}_i = \mathbf{1}_{p_i} = (1,\ldots,1)^T$ and $\mathbf{b}_i = (\alpha_i^C)$. Thus, the covariance structure (6.1.2) reduces to $\sigma_C^2 \mathbf{J}_{p_i} + \sigma^2 \mathbf{I}_{p_i}$, where $\sigma_C^2 = V(\alpha_i^C)$ and assuming that the e_{ij}'s are uncorrelated each with variance σ^2.

The next step up in the hierarchy, to a **three-level** model, may be made by considering children in classes in schools. Then, in (3.1.1), μ_{ij} may now contain child, class and school covariates, and α_{ij} may take the form $\alpha_i^S + \alpha_{i,c(j)}^C$, where α_i^S is the random individual school effect, and $\alpha_{j,c(j)}^C$ is the random individual class effect for child j who is in class $c(j)$. This example is also given by Goldstein (1987, section 2.3). The random term in the model $(\alpha_i^S + \alpha_{i,c(j)}^C + e_{ij})$ covers three levels: e_{ij} for the jth child (level 1), $\alpha_{i,c(j)}^C$ for his class (level 2), and α_i^S for the school (level 3). In the general model (6.1.1) the data vector $\mathbf{y}_i$ now contains the scores for all the tested children in school i, with subvectors representing different classes within the school. The error vector $\mathbf{e}_i$ has components e_{ij}, and the random effects can be represented as $\mathbf{Z}_i\mathbf{b}_i$ with $\mathbf{b}_i = (\alpha_i^S, \alpha_{i1}^C, \ldots, \alpha_{iq}^C)$ when there are q classes, and the jth row of $\mathbf{Z}_i$ having 1's in the first and $c(j)$th positions and 0's elsewhere. Assume, as Goldstein does, that $\mathbf{B} = V(\mathbf{b}_i) = \begin{pmatrix} \sigma_S^2 & 0 \\ 0 & \sigma_C^2 \mathbf{I}_q \end{pmatrix}$, i.e. that the α's are all uncorrelated with the indicated variances. Then, in the covariance structure (6.1.2), $\mathbf{Z}_i\mathbf{B}\mathbf{Z}_i^T$ has the form

$$\begin{pmatrix} (\sigma_S^2 + \sigma_C^2)\mathbf{J}_{11} & \sigma_S^2\mathbf{J}_{12} & \sigma_S^2\mathbf{J}_{13} & \cdots \\ \sigma_S^2\mathbf{J}_{21} & (\sigma_S^2 + \sigma_C^2)\mathbf{J}_{22} & \sigma_S^2\mathbf{J}_{23} & \cdots \\ & \cdot & \cdot & \cdots \end{pmatrix}$$

where the $\mathbf{J}$-matrices have all entries equal to 1, and $\mathbf{J}_{kk}$ is square of order equal to the number of children in class k. For instance, the correlation between classmates is $(\sigma_S^2 + \sigma_C^2)/(\sigma_S^2 + \sigma_C^2 + \sigma^2)$, and that between children in different classes in the same school is $\sigma_S^2/(\sigma_S^2 + \sigma_C^2 + \sigma^2)$.

Goldstein (1987) also considers more elaborate models: in his Chapter 3, Z_i may contain components of X_i, as in growth curve models described above in section 6.3.2; his Chapter 4 covers the case where repeated measures are made on each child or on higher level units; his Chapter 5 considers multivariate data, with more than one response variable, e.g. scores on different subjects.

6.4 Examples

In this section some data sets will be analysed on the basis described in section 6.2.

Example 6.1

The growth curves in Example 2.4 were treated as three groups of individual straight lines. The $p = 11$ occasions are divided into two phases by a change of treatment between occasions 6 and 7. Before applying some of the current techniques let us make some investigations to give some guidance for choice of regression model. In the style of section 4.2 Hotelling's T^2 tests are performed, separately for phases 1 and 2, for equality and parallelism of group mean profiles. Table 6.1 gives the results which more or less confirm the visual impressions from Fig. 2.4. Interpreting the results liberally, for equality Groups 2 and 3 go together in both phases 1 and 2, and for

Table 6.1. *Tests for equality and parallelism of group mean profiles, and for straight-line regression, in Example 6.1*

	Groups	Equality	Parallelism
Phase 1	1 vs. 2	$F_{6,8} = 15 \cdot 8 \ (p < 0 \cdot 001)$	$F_{5,9} = 2 \cdot 56 \ (p > 0 \cdot 10)$
	1 vs. 3	$F_{6,8} = 24 \cdot 8 \ (p < 0 \cdot 001)$	$F_{5,9} = 3 \cdot 67 \ (0 \cdot 025 < p < 0 \cdot 05)$
	2 vs. 3	$F_{6,8} = 1 \cdot 33 \ (p > 0 \cdot 10)$	$F_{5,9} = 0 \cdot 84 \ (p > 0 \cdot 10)$
Phase 2	1 vs. 2	$F_{5,9} = 25 \cdot 7 \ (p < 0 \cdot 001)$	$F_{4,10} = 17.5 \ (p < 0 \cdot 001)$
	1 vs. 3	$F_{5,9} = 33 \cdot 2 \ (p < 0 \cdot 001)$	$F_{4,10} = 13 \cdot 2 \ (p < 0 \cdot 001)$
	2 vs. 3	$F_{5,9} = 4 \cdot 22 \ (0 \cdot 025 < p < 0 \cdot 05)$	$F_{4,10} = 4 \cdot 23 \ (0 \cdot 025 < p < 0 \cdot 05)$

Table 6.2. *Tests for polynomial fits to group mean profiles in Example 6.1*

	Group	Linear	Quadratic
Phase 1	1	$F_{4,10} = 4 \cdot 10$ $(0 \cdot 025 < p < 0 \cdot 05)$	$F_{3,11} = 4 \cdot 52$ $(0 \cdot 025 < p < 0 \cdot 05)$
	2	$F_{4,10} = 0 \cdot 38$ $(p > 0 \cdot 10)$	$F_{3,11} = 0 \cdot 52$ $(p > 0 \cdot 10)$
	3	$F_{4,10} = 2 \cdot 80$ $(0 \cdot 05 < p < 0 \cdot 10)$	$F_{3,11} = 3 \cdot 47$ $(0 \cdot 05 < p < 0 \cdot 10)$
Phase 2	1	$F_{3,11} = 0 \cdot 07$ $(p > 0 \cdot 10)$	$F_{2,12} = 0 \cdot 10$ $(p > 0 \cdot 10)$
	2	$F_{3,11} = 1 \cdot 78$ $(p > 0 \cdot 10)$	$F_{2,12} = 1 \cdot 05$ $(p > 0 \cdot 10)$
	3	$F_{3,11} = 13 \cdot 8$ $(p < 0 \cdot 001)$	$F_{2,12} = 15 \cdot 1$ $(p < 0 \cdot 001)$

parallelism all three groups conform in phase 1 but Group 1 diverges in phase 2. The question of whether straight-line regression is adequate is tackled as described in section 4.3. The results, given in Table 6.2, will allow us to fit straight lines in phase 1 without too much conscience-wrestling.

Let us start with a model of parallel regression lines in phase 1: the mean weight in Group g at time x_j is given by $\mu_j = \alpha_g + \delta x_j$, with slope δ and intercept α_g for Group g ($g = 1, 2, 3$). In the current notation $p = 6$, $q = 4$, $\boldsymbol{\beta} = (\alpha_1, \alpha_2, \alpha_3, \delta)^T$ and $\mathbf{X}_i (6 \times 4)$ has jth row $(1, 0, 0, x_j)$ for Group 1, $(0, 1, 0, x_j)$ for Group 2 and $(0, 0, 1, x_j)$ for Group 3. For illustration of the current approach let us try out various plausible covariance structures:

1. $\boldsymbol{\Sigma}_i = \sigma_1^2 \mathbf{J}_p + \sigma^2 \mathbf{I}_p$: this model arises from random individual levels (with $\mathbf{Z}_i = \mathbf{1}_p$ and $\mathbf{b}_i = (\alpha_i^l)$ in the general scheme) and conditional independence ($\mathbf{E}_i = \sigma^2 \mathbf{I}_p$).
2. $\boldsymbol{\Sigma}_i = \sigma_1^2 \mathbf{J}_p + \mathbf{E}_i$, with Markov-structured $\mathbf{E}_i$;
3. $\boldsymbol{\Sigma}_i = \mathbf{Z}_i \mathbf{B} \mathbf{Z}_i^T + \sigma^2 \mathbf{I}_p$, $\mathbf{Z}_i (6 \times 2)$ having jth row $(1, x_j)$ and $\mathbf{B}$ (2×2) positive semi-definite: this model arises from random individual regression lines (with coefficients $\mathbf{b}_i^T = (\alpha_i^l, \delta_i^l)$ and $\mathbf{B} = \begin{pmatrix} \sigma_\alpha^2 & \sigma_{\alpha\delta} \\ \sigma_{\alpha\delta} & \sigma_\delta^2 \end{pmatrix}$ in the general scheme), and then $\mathbf{Z}_i \mathbf{B} \mathbf{Z}_i^T$ (6×6) has (j, k)th entry $\sigma_\alpha^2 + (x_j + x_k)\sigma_{\alpha\delta} + x_j x_k \sigma_\delta^2$.
4. $\boldsymbol{\Sigma}_i = \mathbf{Z}_i \mathbf{B} \mathbf{Z}_i^T + \mathbf{E}_i$ with Markov $\mathbf{E}_i$.

The log-likelihood ratio test between models (1) and (2) yields $\chi_1^2 = 80 \cdot 59$ ($p < 0 \cdot 0005$), and that between models (3) and (4) yields

$\chi_1^2 = 4.06 \, (0.025 < p < 0.05)$. Thus it seems that with the simpler form of $\mathbf{b}_i$-variation the rules imply that we must allow $\rho \neq 0$ in the error structure. However, the rules were drawn up on the assumption that the model is correct, and the model rests on a lot of assumptions. For instance, fitting a straight line to points which follow a shallow curve will produce residuals with high positive autocorrelation. The comparison between models (2) and (4) yields $\chi_2^2 = 2.50 \, (p > 0.20)$ so it seems that model (2) is adequate. Note that model (2) is the parametric restriction of model (4) obtained by taking $\sigma_\delta = 0$ (and therefore $\sigma_{\alpha\delta} = 0$, also). Some numerical results from model (2) are $\hat{\boldsymbol{\phi}} = (\hat{\sigma}, \hat{\rho}, \hat{\sigma}_1) = (21.62, \, 0.9954, \, 26.36)$, $(\hat{\alpha}_1, \hat{\alpha}_2, \hat{\alpha}_3) = (247, 460, 508)$ with standard errors $(12.0, \, 16.9, \, 16.9)$, and $\hat{\delta} = 0.602$ with standard error 0.084. □

The tests in Example 6.1 were performed in the standard manner, referring twice the log-likelihood difference for nested models to a chi-square distribution. However, recent work on tests for parameters at boundaries (e.g. Self and Liang, 1987; Miller, 1977) suggests that the chi-square may provide a poor approximation even to the large-sample distribution. This digression will not be pursued here.

Example 6.2

The data, with $n = 12$ patients and $p = 7$ occasions in three phases, were subjected to Anova in Example 3.3 and to Hotelling's T^2 in Example 4.2. The evidence was strongly in favour of differences in level between phases not within phases. The corresponding regression model was fitted in Example 5.1 with unstructured covariance. The exercise will now be repeated but with a structured form for $\boldsymbol{\Sigma}$.

The Markov correlation model is plausible for $\mathbf{E}_i$ since the measurements are made at unequally spaced times over a period of 16 weeks. For $\mathbf{B}$ adoption of the general form $\mathbf{A}\mathbf{A}^T$, with $\mathbf{A}$ (3×3) lower triangular, only costs 6 parameters. Likelihood ratio tests between the four covariance models, as in Example 6.1 above, point towards adoption of $\rho \neq 0$ very strongly, but do not support the form $\mathbf{X}_i\mathbf{A}\mathbf{A}^T\mathbf{X}_i^T$ over $\sigma_1^2\mathbf{J}_p$ strongly ($\chi_5^2 = 10.30, \, 0.05 < p < 0.10$). On the basis of the latter form the estimates are $(\hat{\sigma}, \hat{\rho}, \hat{\sigma}_1) = (0.288, \, 0.462, \, 0.122)$, and $\hat{\boldsymbol{\beta}}^T = (\hat{\alpha}_1, \hat{\alpha}_2, \hat{\alpha}_3) = (0.530, \, 1.14, \, 0.701)$ with estimated covariance matrix

$$\hat{\mathbf{V}}_\beta = 10^{-3} \begin{pmatrix} 5\cdot94 & \cdot & \cdot \\ 2\cdot58 & 4\cdot97 & \cdot \\ 1\cdot73 & 2\cdot58 & 5\cdot94 \end{pmatrix}.$$

The estimates for $\boldsymbol{\beta}$ and $\mathbf{V}_\beta$ here bear a rough resemblance to those in Example 5.1.   □

Example 6.3

The hip-replacement data, with $n = 30$ patients and $p = 3$ occasions, were analysed in Example 5.4. The fitted linear model contained two dummy indicators, for three occasions, plus age and sex covariates. The same linear model will be fitted now. In contrast, the covariance matrix will now be structured as $\sigma_1^2 \mathbf{J}_p + \mathbf{E}_i$, as in models (1) and (2) of Example 6·1. The likelihood ratio test for $\rho = 0$ yields $\chi_1^2 = 0\cdot17$ so model (1) seems adequate. The numerical results are $(\hat{\sigma}, \hat{\sigma}_1) = (3\cdot71, 1\cdot56)$ and $\hat{\boldsymbol{\beta}} = (38\cdot2,\ 28\cdot4,\ 30\cdot8,\ 0\cdot921,\ 0\cdot034)^{\mathrm{T}}$ with estimated covariance matrix very close to that given previously.   □

6.5   Further reading

The model (6.1.1) is given by Laird and Ware (1982) following Harville (1977) and earlier writers. These papers cover important aspects not included in the present chapter such as REML (restricted maximum likelihood) and the Bayesian approach.

Rao's (1965) work deals with the case where $n = 1$, $\mathbf{Z}_i = \mathbf{X}_i$ and there is an independent estimate of $\boldsymbol{\Sigma}$ available. In Rao (1967) $n > 1$, $\mathbf{Z}_i = \mathbf{X}_i$ and attention is directed to efficiency of $\hat{\boldsymbol{\beta}}$, estimation of $\mathbf{b}_i$ and prediction for an individual. Rosenberg (1973) and Rao (1975) cover similar ground.

Swamy (1970) looked at the case $p_i = p$, $\mathbf{E}_i = \mathrm{diag}(\sigma_1^2, \ldots, \sigma_p^2)$. He proved consistency of $\hat{\boldsymbol{\beta}}$ together with that of certain unbiased estimators for $\mathbf{B}$ and $\mathbf{E}_i$ which are not maximum likelihood. Carter and Yang (1986) point out that Swamy's $\hat{\mathbf{B}}$ can be nonpositive-semi-definite and suggest a cure, also non-m.l.e.

Lindley and Smith (1972) add a third stage to the two-stage model to incorporate Bayesian prior distributions for the parameters. Fearn (1975) develops this for specific application to growth curves.

Reinsel (1982, 1984) extends the growth curve set-up to multivariate repeated measures where more than one characteristic is recorded at each measurement occasion. A similar extension is made by Schaeffer, Wilton and Thompson (1978).

Section 6.2 is based in part on Searle (1971, section 10.8) after Henderson *et al.* (1959), and Laird and Ware (1982) after Harville (1976).

Diggle (1988) makes a case for general adoption of $\Sigma_i = \sigma^2 \mathbf{I}_p + \sigma_\alpha^2 \mathbf{J}_p + \sigma_R^2 \mathbf{R}_i$; here, $\sigma^2 \mathbf{I}_p + \sigma_\alpha^2 \mathbf{J}_p$, arises from 'error + random individual levels' as in section 3.1, and $\mathbf{R}_i$ is modelled as $(\mathbf{R}_i)_{jk} = \exp(-\rho|t_j - t_k|^c)$ with $c = 1$ or 2 ($c = 1$ being equivalent to the form given above in section 5.4.3). Diggle gives a general discussion of modelling in this context, and goes on to describe likelihood-based inference and the choice and validation of the $\mathbf{R}_i$-structure using the empirical semi-variogram. Finally, he applies the methods to two examples.

Computational aspects have received a fair degree of coverage. The EM algorithm has been used by Dempster, Rubin and Tsutakawa (1981) and by Laird, Lange and Stram (1987). An alternative method is proposed by Longford (1987).

CHAPTER 7

Crossover experiments

Crossover designs and their analysis are an extensive topic in their own right. In a short chapter in a book such as this we can do no more than introduce them. It can be argued that crossover designs are quite distinct from repeated measurements (Yates, 1982), but for the present account we have taken a wider definition.

7.1 Simple 2 × 2 designs

The most basic form of the crossover design involves measuring a group of subjects first under condition A and secondly under condition B, and measuring a separate group of subjects first under B and then under A. Such designs have been used extensively in clinical trials but have suffered criticism from the regulatory bodies of the pharmaceutical industry for reasons explained below.

Typically, conditions A and B are two different treatments, the primary objective being to compare them. The virtue of such designs is that the A–B comparison is made within subjects. Thus if, as is often the case, there is considerably more variation between subjects than within subjects, then a more powerful test of treatment differences will result than if one had simply compared two independent groups. In effect, each subject acts as his own control.

Since the conditions are presented sequentially to each subject, there are some obvious constraints on the type of study for which the method is appropriate. For example, in a medical situation one could not apply a crossover design if A and B were being tested as potential curative agents. In other application areas one could not apply such a design if A and B were potentially destructive. In general one requires the units to revert to their original state after experiencing the first condition (but see the discussions of period effect below).

The basic design is

	Period 1	Period 2
Group 1	Treatment A	Treatment B
Group 2	Treatment B	Treatment A

with subjects being randomly allocated to each of the two groups.

We shall suppose initially that there is a treatment and period effect, but no interaction between treatment and period – that is, the results are not influenced by the order of presentation of the two treatments.

Equation (3.1.1) presented a fundamental model for repeated measures designs: $y_{ij} = \mu_{ij} + \alpha_{ij} + e_{ij}$, i referring to subject and j to time. It will be more convenient here to make the group membership explicit, writing.

$$y_{igj} = \mu_{gj} + \alpha_{igj} + e_{igj} \tag{7.1.1}$$

for group $g = 1, 2$; subject $i = 1, \ldots, n_g$; period $j = 1, 2$, where n_g is the number of subjects in group g. Initially we shall take

$$\mu_{gj} = \mu + T_{gj} + P_j \tag{7.1.2}$$

where $T_{gj} = T_A$ (if $g = j$) or T_B (if $g \neq j$), T_A and T_B being the effects of treatments A and B respectively and P_j being the effect of period j. In the above $\alpha_{igj} = \alpha_{ig}$ is the random effect for subject i in group g, and we shall assume that $\alpha_{ig} \sim N(0, \sigma_\alpha^2)$, independent for all i, g. Similarly, e_{igj} is measurement error, $e_{igj} \sim N(0, \sigma^2)$ independent for all i, g, j.

To see where the power of this design comes from we work within subjects and take difference scores. For simplicity at this stage we shall assume $n_1 = n_2 = n$. Then

$$d_{i1} = y_{i11} - y_{i12} = (T_A - T_B) + (P_1 - P_2) + (e_{i11} - e_{i12}),$$
$$d_{i2} = y_{i22} - y_{i21} = (T_A - T_B) + (P_2 - P_1) + (e_{i22} - e_{i21}).$$

From the distributional assumptions on e_{igj},

$$(\bar{d}_{\cdot 1} + \bar{d}_{\cdot 2})/2 \sim N(T_A - T_B, \sigma^2/n)$$

where $\bar{d}_{\cdot g}$ is the mean of the d_{ig} scores in group g. In contrast, a simple two-independent-groups test, with n subjects in each group, has

$$(\bar{y}_{\cdot 11} - \bar{y}_{\cdot 21}) \sim N(T_A - T_B, 2(\sigma_\alpha^2 + \sigma^2)/n).$$

Even if we double the number of patients, so that the same numbers of measurements are being made in the two cases the variance of this second estimate is $(\sigma_\alpha^2 + \sigma^2)/n$, inflated relative to the crossover approach by the between-subjects variance.

In the crossover approach we can estimate the period effect by

$$(\bar{d}_{.1} - \bar{d}_{.2})/2 \sim N(P_1 - P_2, \sigma^2/n).$$

To conduct tests based on these distributions we will, of course, estimate the variances of $(\bar{d}_{.1} \pm \bar{d}_{.2})/2$ directly from the d_{ig}. If the groups are unequal the only complication is that the variances are estimated by $\frac{1}{4}(\sigma_D^2/n_1 + \sigma_D^2/n_2)$ where σ_D^2 is the (assumed common) variance of the difference scores in each group.

Thus far we have assumed that there is no *treatment × period* interaction – that the results are not affected by the order in which the two treatments are given. We now relax this requirement and explore how one might test it. We first note that in the simple design under discussion the *treatment × period* interaction is completely confounded with the *group* main effect (and that the *group × period* interaction is confounded with the *treatment* main effect, and that the *group × treatment* interaction is confounded with the *period* main effect). But we also note that in a randomized design it would be reasonable to assume no group effect.

Interactions can arise for several reasons. Some external influence may change over time, so that the conditions under which the two treatments are given differ between periods 1 and 2. Measurement scale peculiarities can also introduce an apparent interaction: subjects with high (low) initial scores may experience large (small) differences between treatments. Conversely, 'ceiling' or 'floor' effects (i.e. natural upper or lower bounds on the observed values) may distort otherwise straightforward additivity.

For 2 × 2 designs a **residual** or **carryover** effect is indistinguishable from a *treatment × period* interaction. Here a residual effect occurs when treatment A in period 1 produces an effect in period 2, and treatment B in period 1 produces an effect in period 2, and these effects differ. (If they do not differ then the effect is indistinguishable from a period effect.) **Washout periods** between treatments can sometimes be used to eliminate residual effects, but not always. One troublesome type of residual effect is the psychological one. If A is effective and B is not, then subjects are likely to approach the second period in different frames of mind, according to which treatment they have already received.

The parametrization for interaction (*treatment* × *period*) that we shall adopt is

$$\mu_{gj} = \mu + T_{gj} + P_j + (j-1)I_g \qquad (7.1.3)$$

Note that I_g only enters the model explicitly in period 2 ($j = 2$); when $I_1 = I_2$ there is, of course, no interaction. To test for interaction we define

$$m_{ig} = y_{ig1} + y_{ig2} = 2\mu + T_A + T_B + P_1 + P_2 + I_g + 2\alpha_{ig} + e_{ig1} + e_{ig2}$$

for $g = 1, 2$. Then $\bar{m}_{.1} - \bar{m}_{.2} \sim N(I_1 - I_2, (8\sigma_\alpha^2 + 4\sigma^2)/n)$. In practical testing one estimates the variances of $\bar{m}_{.1}$ and $\bar{m}_{.2}$ directly from the m_{ig} so that the variance of $(\bar{m}_{.1} - \bar{m}_{.2})$ is $\sigma_s^2/n_1 + \sigma_s^2/n_2$ where σ_s^2 is the variance (assumed the same for the two groups) of $(y_{ig1} + y_{ig2})$.

We can now see where the anxiety about the simple crossover design comes from. In the earlier test for a treatment effect, assuming no *treatment* × *period* interaction, the variance of the estimator was σ^2/n. In the test for interaction it is $(8\sigma_\alpha^2 + 4\sigma^2)/n$. Thus the interaction test will be much less powerful; it will be more likely to fail to detect sizeable interactions, so that bias is introduced into the consequent estimate of the treatment main effect. In fact, if I_1 and I_2 are incorrectly assumed to be equal then the test statistic for treatment effect is distributed as

$$(\bar{d}_{.1} + \bar{d}_{.2})/2 \sim N((T_A - T_B) + (I_2 - I_1)/2, \sigma^2/n)$$

so that the F-statistic for a treatment effect will not follow a central F-distribution under the null hypothesis if $I_2 \neq I_1$.

Should the interaction test, insensitive though it is, produce a significant result, then one will need to test the treatment effect making allowance for this. The appropriate approach in such a case is to use merely the responses in period 1, since up to this point (assuming random allocation) the subjects have been treated equivalently. This test, however, is based on a variance $2(\sigma_\alpha^2 + \sigma^2)/n$, much larger than that for the crossover test. Moreover, since the sample sizes will have been chosen with the more powerful crossover in mind, it will probably be the case that the single period test will not be sufficiently powerful to be of great value. Neither is the problem eased by increasing sample size to make the interaction test more powerful since as sample size increases so the treatment effect also gains sensitivity to smaller interaction effects. The conclusion seems to be that the simple crossover design, with the analysis presented above, should not be used.

Example 7.1

Despite the weaknesses of the design outlined above, it is still used quite often. To illustrate the calculations we have taken the data analysed by Hills and Armitage (1979), reproduced here in Table 7.1. The experiment was a comparison of a drug to treat enuresis with a placebo. Treatment A consisted of 14 consecutive days using the drug, and treatment B of 14 consecutive days using the placebo. The response variable was the number of dry nights out of 14.

From the discussion above, a statistic for the treatment effect is $(\bar{d}_{.1} + \bar{d}_{.2})/2 = 2\cdot03$ with estimated standard error 0·62, giving a 95% confidence interval from 0·76 to 3·30 days.

For the *treatment × period* interaction, we have $(\bar{m}_{.1} - \bar{m}_{.2}) = 16\cdot58 - 13\cdot41 = 3\cdot17$ and the variance of $(\bar{m}_{.1} - \bar{m}_{.2})$ is estimated to be 41·9; t-tables with $(17 + 12 - 2) = 27$ d.f. show this to be nonsignificant.   □

Table 7.1. *Simple crossover design*

Treatment order	A–B		B–A	
Period	1	2	1	2
	8	5	12	11
	14	10	6	8
	8	0	13	9
	9	7	8	8
	11	6	8	9
	3	5	4	8
	6	0	8	14
	0	0	2	4
	13	12	8	13
	10	2	9	7
	7	5	7	10
	13	13	7	6
	8	10		
	7	7		
	9	0		
	10	6		
	2	2		

Data from Hills and Armitage (1979).

In the preceding discussion part of the problem has arisen because of the between-subjects variance entering as additional error variance in some tests. One way in which this term can be removed is by taking an initial baseline score for each subject, before the first treatment period, and then working with differences from this baseline. Of course, there may be practical implications since now three measurements need to be taken. This leads, under the same assumptions as before, to

$$(\bar{d}_{\cdot 1} + \bar{d}_{\cdot 2})/2 \sim \mathrm{N}((T_A - T_B) + (I_2 - I_1)/2, \sigma^2/n)$$

and

$$\bar{m}_{\cdot 1} - \bar{m}_{\cdot 2} \sim \mathrm{N}(I_1 - I_2, 12\sigma^2/n)$$

where now $\bar{d}_{\cdot g}$ and $\bar{m}_{\cdot g}$ refer to mean differences from baseline. We see that even eliminating the between-subjects variation may not improve things – in the expressions for variance, σ_α^2 is transformed to σ^2 rather than being removed altogether.

7.2 A Bayesian approach to 2 × 2 designs

Grieve (1985) considered the simple design with no baselines, with the parametrization as above and with $T_A = -T_B = T$, $P_1 = -P_2 = P$, $I_1 = -I_2 = I$; also, in his notation $\sigma_A^2 = 2\sigma_\alpha^2 + \sigma^2$, $N = n_1 + n_2$, and $m = N/n_1 n_2$.

The uninformative prior $P(\mu, P, T, I, \sigma^2, \sigma_A^2) \propto \sigma^{-2}\sigma_A^{-2}$ yields posteriors

$$P(I|y) = t(\hat{I}, mS_\alpha/2(N-2), N-2),$$
$$p(T|I, y) = t(\hat{T} + \hat{I}/2, mS/8(N-2), N-2),$$
$$p(T|y) \simeq t(\hat{T} + \hat{I}/2, mh/8f, f)$$

where $t(\theta, \phi, \nu)$ denotes a shifted and scaled t-distribution with ν degrees of freedom, location parameter θ, and scale parameter $\phi^{1/2}$. Here

$$\hat{I} = \tfrac{1}{2}(\bar{y}_{\cdot 11} + \bar{y}_{\cdot 12} - \bar{y}_{\cdot 21} - \bar{y}_{\cdot 22}),$$
$$\hat{T} = \tfrac{1}{4}(\bar{y}_{\cdot 11} + \bar{y}_{\cdot 12} - \bar{y}_{\cdot 21} + \bar{y}_{\cdot 22}),$$
$$S_\alpha = 2\left(\sum_{ig} \bar{y}_{ig\cdot}^2 - \sum_g n_g \bar{y}_{ig\cdot}^2\right),$$
$$S = \sum_{igj} y_{igj} - S_\alpha - \sum_g n_g \sum_k \bar{y}_{\cdot gk}^2,$$
$$f = \{(S + S_\alpha)^2(N-6)/(S^2 + S_\alpha^2)\} + 4,$$
$$h = (f-2)(S + S_\alpha)/(N-4).$$

Using these posterior distributions on the data in Table 7.1 shows the effect of a nonzero interaction on the treatment effect: $P(T > 0 | I = 0, y) = 0.999$ while $P(T > 0 | y) = 0.627$.

Forms for the posterior distributions conditional on the additional constraint that $\sigma_A^2 > \sigma^2$ are also given in Grieve (1985), but he notes that in general for two-period crossover designs conditioning on this constraint has very little effect.

To combine the two situations of interaction/no interaction into a single posterior, one approach is to use a Bayes factor. This yields

$$p(T | y) = \{\pi B/(1 + \pi B)\} p(T | y, M_0) + \{1/(1 + \pi B)\} p(T | y, M_1)$$

where M_0 is the model without a carryover effect, M_1 is the model with carryover, π is the prior odds on M_0, and B is the Bayes' factor,

$$B = (3/2m)^{1/2} \{1 + F/(N - 2)\}^{-N/2}$$

where F is the usual F-statistic for assessing the significance of an interaction.

In discussing the Bayesian approach to crossover designs, Racine-Poon *et al.* (1986) conclude: 'For in-house analysis, we believe there to be great potential for these kinds of rich and flexible analysis displays. For external reporting purposes, our experience thus far with this approach to the analysis of crossover trials suggests that the "unequivocal evidence" of no carryover effect will rarely be forthcoming.'

7.3 More complex crossover designs for two treatments

Up to now we have restricted our discussion to the case of the simplest of two-treatment designs, i.e. those with only two periods, just touching on the effect of adding baseline measurements. The difficulty with these designs is, as we have seen, the problem of disentangling residual or *treatment × period* interaction effects from the treatment effect. We now turn to consideration of more complex crossover designs which will permit this. We shall, however, continue to restrict our discussion to studies involving only two treatments.

We note here that increasing the number of periods may introduce practical problems. At least it necessarily increases the length of the trial, which may mean a loss of subjects completing all periods. With this practical aspect in mind, we shall restrict our generalization of 2×2 designs to designs involving just three periods. We shall also consider only the two-group case, i.e. with only two distinct treatment sequences.

In cases where more than two periods are involved it is necessary to be more precise about residual effects of treatments: there will be a first-order residual effect, which lasts only to the period following the treatment, a second-order residual effect, which lasts only through the two periods following the treatment, and so on. Each of these may be an object of investigation, and each may have to be eliminated in some way when estimating or testing the others. An associated concept is the cumulative treatment effect. This is the arithmetic sum of the direct treatment effect and all residual treatment effects.

The model we shall adopt is, as before

$$\mu_{gj} = \mu + T_{gj} + P_j + R_r \qquad (7.3.1)$$

where T_{gj} is the treatment, T_A or T_B depending upon the group and period according to the particular design being used, P_j is the period effect and R_r is the first-order residual effect, zero for the first period and R_A or R_B according to the preceding treatment. Note that we are here assuming no second-order residual effect and no additional interaction terms.

The treatment effect $(T_A - T_B)$ will be estimated by a weighted combination of the mean scores in the *group × period* cells, with the weights depending on the particular design and on the number of cases per group. For simplicity we shall again assume equal numbers (n) in each group. Our objective, as before, is to derive an estimate of $(T_A - T_B)$ using contrasts over periods (rather than groups) so that the smaller within-subjects variance applies. Then, if u_{gj} is the weight for the gth group at the jth period and σ^2 is the measurement error, as in section 7.1, the variance of the estimate of treatment effect will be $(\sigma^2/n)\sum_g\sum_j u_{gj}^2$.

The residual effect $(R_A - R_B)$ will be estimated in the same way, with weights v_{gj}, leading to variance $(\sigma^2/n)\sum_g\sum_j v_{gj}^2$. In both expressions the design influences the variance only through the second factor, so that we may choose between designs merely by selecting that with the smallest sum of squared weights.

The possibility of adding baselines, mentioned above, doubles the number of designs we shall consider. We assume in the results below that each treatment period measurement has been preceded by a baseline measurement. Again, of course, we should note the possible

Table 7.2. *Some more complicated designs for two treatments*

	Design				Coeff. matrix for T effect			Coeff. matrix for R effect			SS weights without baseline		SS weights with baseline	
											T	R	T	R
		Period												
		1	2	3										
Group	1	A	B	–	1	0	–	1	1	–	–*	–	4·00	8·00
	2	B	A	–	−1	0	–	−1	−1	–				
	1	A	R	B	$\frac{1}{2}$	0	$-\frac{1}{2}$	$-\frac{1}{2}$	1	$-\frac{1}{2}$	2·00	6·00	2·00	3·00
	2	B	R	A	$-\frac{1}{2}$	0	$\frac{1}{2}$	$\frac{1}{2}$	−1	$\frac{1}{2}$				
	1	A	B	R	$\frac{2}{3}$	$-\frac{1}{3}$	$-\frac{1}{3}$	$\frac{1}{3}$	$\frac{1}{3}$	$-\frac{2}{3}$	2·67	2·67	2·29	1·14
	2	B	A	R	$-\frac{2}{3}$	$\frac{1}{3}$	$\frac{1}{3}$	$-\frac{1}{3}$	$-\frac{1}{3}$	$\frac{2}{3}$				
	1	A	B	B	$\frac{1}{2}$	$-\frac{1}{4}$	$-\frac{1}{4}$	0	$\frac{1}{2}$	$-\frac{1}{2}$	1·50	2·00	1·41	1·00
	2	B	A	A	$-\frac{1}{2}$	$\frac{1}{4}$	$\frac{1}{4}$	0	$-\frac{1}{2}$	$\frac{1}{2}$				
	1	A	A	B	0	$\frac{1}{2}$	$-\frac{1}{2}$	−1	1	0	2·00	8·00	1·60	3·40
	2	B	B	A	0	$-\frac{1}{2}$	$\frac{1}{2}$	1	−1	0				
	1	A	B	A	1	$-\frac{1}{2}$	$-\frac{1}{2}$	1	0	−1	6·00	8·00	2·18	1·54
	2	B	A	B	−1	$\frac{1}{2}$	$\frac{1}{2}$	−1	0	1				

*The treatment and residual effects cannot be estimated using within-groups contrasts

adverse practical implications of increasing the number of measurements, even if it leads to greater estimation efficiency.

Table 7.2 shows some appropriate designs which might be considered. The table shows the coefficient matrices for calculating the treatment and residual effects, and the resulting sums of squared weights for the two cases with and without baselines. In this table A and B signify treatments A and B, and R signifies a rest period – a period during which no treatment is given, but a measurement is still taken.

A glance at the sum-of-squares columns leads to the recommendation to use the ABB/BAA design if it is ethically possible to submit the subjects to three treatment exposures.

7.4 Crossover trials with a binary response

When the response is binary (1/0, success/failure, good/bad, etc.) one will be seeking to estimate the effects of treatment (and perhaps period, etc.) on the probability of scoring 1 or 0. Again we shall consider the 2×2 design.

We first introduce some notation. There are four possible score profiles for each individual, namely 00, 01, 10 and 11 where ij refers to score i in the first period and j in the second. Of these, only 01 and 10 imply a difference between treatments. Grouping the numbers who score 00 or 11 (i.e. no difference) together, we have the following table.

Score profile	Group 1 (Sequence AB)	Group 2 (Sequence BA)
01	n_{11}	n_{12}
00 or 11	n_{21}	n_{22}
10	n_{31}	n_{32}

A straightforward test for the hypothesis of no *treatment* effect, in a model which assumes no *period* effect or *treatment × period* interaction, is given by a binomial test comparing $(n_{12} + n_{31})/(n_{11} + n_{32} + n_{12} + n_{31})$ with $1/2$. This is the McNemar test, and simply tests whether the proportion who score 1 on treatment A, among those whose scores differ between A and B, is $1/2$. This test will also be legitimate as a test of treatment effect in the presence of a period effect (but no interaction) if the two groups have equal numbers of subjects.

More generally, suppose that there is a *period* effect but no *treatment × period* interaction, and that the null hypothesis H_0 is one of no *treatment* effect. Then, under H_0, the proportion preferring period 1 (i.e. with score profile 10) should be the same for the two groups, and similarly the proportion preferring period 2 (with score profile 01) should be the same for the two groups. Thus a χ^2 test on the 3×2 table indicated above would be appropriate.

More sensitive tests result, however, if one focuses attention on the particular patterns of departure from independence that one would expect if there were a *treatment* effect. The Mainland–Gart test focuses on the four cells in which a difference occurs by applying Fisher's exact test to the 2×2 table

$$\begin{pmatrix} n_{11} & n_{12} \\ n_{31} & n_{32} \end{pmatrix}.$$

An alternative due to Prescott (1981) is to include the 'no preference' groups by recoding a score profile 01 as -1, profiles 00 and 11 as 0, and profile 10 as $+1$. Then if, for example, A has a higher success rate than B, group 1 will have a higher mean score than group 2. This can be tested by a randomization t-test. In fact the test statistic can be simplified to $T = n_{11} - n_{31}$ which, under the null hypothesis, has expectation estimated by $n_{.1}(n_{1.} - n_{3.})n_{..}$, variance estimated by

$$n_{.1}n_{.2}\{(n_{1.} + n_{3.}) - (n_{1.} - n_{3.})^2/n_{..}\}/\{n_{..}(n_{..} - 1)\}$$

and is asymptotically normally distributed; thus one may use $z = (|T - E(T)| - 1/2)/\sqrt{\text{var}(T)}$ as a large-sample test statistic. For small samples a one-tailed test against the alternative that A is more successful than B is given by regarding the row and column totals as fixed and calculating

$$P = \sum_{i=T^*}^{n_{1.}} \sum_{j=0}^{i-T^*} \binom{n_{1.}}{i}\binom{n_{2.}}{n_{.1}-i-j}\binom{n_{3.}}{j} \Big/ \binom{n_{..}}{n_{.1}}$$

which is the probability under H_0 of observing a configuration with $T > T^*$, the observed value of T.

The Mainland–Gart test does not require random allocation of subjects to the two groups and thus could be applied even if purposive allocation had been adopted. In contrast, Prescott's test does assume random allocation. In fact, of course, the latter will be the norm in a clinical trial. Prescott (1981) has conducted simulations which suggest that his test is more powerful than the Mainland–Gart test if there is a period effect (and than the McNemar test with equal group sizes and a period effect). Fidler (1984) has explored Prescott's results analytically and concluded that if both group and carryover effects are absent then Prescott's test is better than the Mainland–Gart test, supporting Prescott's simulation results. One should, of course, recall that one aim of randomization is to eliminate group effect. One could rely on a pre-test of carryover before proceeding to test treatment effect but, as in the continuous case, this may not be very powerful.

7.5 Further topics

In this brief review chapter we have made no attempt to cover designs involving more than two treatments. Although two-treatment designs are by far the most popular, designs with more than two are used, and a large number of such designs are possible. One additional possibility, beyond that available for two-treatment designs, is that in multiple-treatment designs one has the flexibility to choose which of the between-treatment contrasts should be the most accurately estimated. For example, if one is comparing several new drugs with a standard treatment the comparisons between the new drugs may be of less intrinsic interest than the comparisons with the standard. The implication is that one must be clear precisely what are one's principal interests before attempting to choose a design. An introduction to the literature of multiple-treatment designs is given in Chapter 5 of Jones and Kenward (1989).

A second topic that we have not discussed is how to handle the case when the measurements are repeated within each time period. That is, a sequence of measurements may be taken while the subject experiences treatment A in period 1, and so on. Such a situation is extremely common in clinical trials. A relatively straightforward approach is based on response feature analysis (Chapter 2): summarize the sequence of observations for each subject in each period by a relevant statistic (e.g. the mean score, AUC, etc.) and use these summary values in a standard crossover analysis. An alternative approach is to calculate the effect of interest (e.g. treatment effect) for each of the p repeated measures in turn (thus: combine the first measures from each period to give an estimate of the mean effect, then combine the second measures in the same way, and so on). This yields p estimates of the effect of interest and these can be analysed by the repeated measures analysis techniques described elsewhere in this book. This is repeated for each effect of interest. A good reference here is Dunsmore (1981b). He compares two approaches, those of Wallenstein and Fisher (1977) and an adaptation of Fearn (1975, 1977), to some data with six repeated measures in each of two periods with 14 heart disease patients. The methods are discussed and the numerical results are examined.

Recently a general log-linear modelling approach to crossover designs with binary data has been developed (Kenward and Jones, 1987), which permits straightforward generalization to more complicated designs (see for example, Jones and Kenward, 1987). For each group they model the possible profiles (00, 01, 10, 11 for the 2×2 case) as a multinomial distribution generated from log probabilities defined according to a linear function of the parameters.

Since the groups are independent this produces overall a product multinomial distribution. This is complicated, but correct statistics using a package such as GLIM can be obtained if the data are assumed to be distributed as (*number of groups*) × (*number of profiles*) independent Poisson variates.

7.6 Further reading

For the simple 2×2 design, Willan and Pater (1986) explored the size of interaction that is necessary in order for the test based solely on the first period to be more powerful than the straightforward crossover test. They found that the crossover test is more powerful when

$$(I_1 - I_2)/(T_A - T_B) < 2 - \{2\sigma^2/(\sigma_\alpha^2 + \sigma^2)\}^{1/2}$$

and concluded that this is often likely to be the case. They also showed that the mean squared error of the crossover estimate of treatment is less than the mean squared error of the estimate of treatment effect based on the first period alone if and only if

$$(I_1 - I_2)^2 < 4(2\sigma_\alpha^2 + \sigma^2)/n.$$

Brown (1980), Willan and Pater (1986), and Freeman (1987) all argue against the procedure of Grizzle (1965) of conducting an initial test of interaction and then following the crossover analysis or the first period analysis according to the results. Freeman also points out that the type I error for this procedure is not the nominal value. With $(I_1 - I_2)n^{1/2}/\sigma$ around 5 it is 0·44, and even with $I_1 = I_2$ it is 0·08.

Bayesian approaches are treated further by Racine *et al.* (1986), who also consider informative priors.

On more complex crossover designs for two treatments, Kershner and Federer (1981) consider designs involving up to six groups and four periods. They also discuss the estimation of other contrasts (e.g. second-order residual effects) and the possibility of optimal allocation of sampling units to reduce the variances of selected estimators, and present results similar to the above for estimating the cumulative treatment effect. Bishop and Jones (1984) review higher-order crossover designs. Matthews (1987) generates optimal two-treatment crossover designs for the case of autocorrelated errors.

For binary responses, the test of treatment effect in the 2×2 case in the presence of a period effect outlined above was proposed by

Mainland (1963) and the theory was developed by Gart (1969). Farewell (1985) gives further discussion of crossover designs for binary data.

Finally, readers requiring an extensive discussion of the whole field can do no better than refer to the excellent book by Jones and Kenward (1989).

CHAPTER 8

Categorical data

8.1 Introduction

Categorical data appear often in repeated measurements problems. A simple example is a preference score measured at several time points. For example, at each of p times the respondents may be asked to express a preference between two alternatives A and B according to the following scheme:

1 = strong preference for A
2 = mild preference for A
3 = no preference
4 = mild preference for B
5 = strong preference for B

A more complicated example is given by the Karnofsky rating scale (Karnofsky and Burchenal, 1949), which provides a measure of ability to cope with everyday activities. It is used, for example, in brain tumour studies. It has eleven categories, scored as follows:

0 = dead
1 = moribund, fatal processes progressing rapidly
2 = hospitalization necessary, very sick, active supportive treatment necessary
3 = severely disabled, hospitalization indicated although death not imminent
4 = disabled, requires special care and assistance
5 = requires considerable assistance and frequent medical care
6 = requires occasional assistance but is able to care for most needs
7 = cares for self, unable to carry on normal activity or to do active work
8 = normal activity with effort, some signs or symptoms of disease
9 = able to carry on normal activity, minor signs or symptoms of disease
10 = normal, no complaints, no evidence of disease

These two examples are ordinal. Undoubtedly the most common way of handling such data is to regard the scores as being on interval scales. This option is, of course, not available for nominal data. So, for example, in analysing voting behaviour over time, where the response categories are three political parties, one could not use such a method and those described below would have to be adopted.

As is to be expected, situations with two response categories are especially common. The categories may be yes/no, success/failure, above/below (a threshold), and so on.

In repeated measurements situations, with individuals moving from one category to another over the measurement occasions, it will sometimes be reasonable to assume that the process is stationary. That is, the probability of moving from category a to category b from time j to $j + 1$ is the same for all $j = 1, \ldots, p - 1$. This may be reasonable in some cases (e.g. some econometric studies) but in others (e.g. clinical trials) it would not normally be an appropriate assumption. The next two sections explore the notion of stationarity in more detail.

Methods for comparing behaviour at different times and between groups of subjects, as well as methods for fitting more complicated models, are outlined in sections 8.4 and 8.5. In section 8.4 a general linear model approach is described. This is based on the multinomial distribution of the subjects' sequences of responses over the set of all possible sequences of responses. In section 8.5 an alternative approach using randomization tests to compare the distributions of responses at each time is outlined. Section 8.5 concludes with a comparative discussion of the two approaches.

8.2 Markov chain models

The most straightforward case arises when the patterns of change in individuals are known. Suppose that n subjects are each observed at p times and let the **profile record** for the ith subject be $\mathbf{y}_i = (y_{i1}, \ldots, y_{ip})^{\mathrm{T}}$, where y_{ij} indicates the **category** (or **state**) in which subject i was observed at time j. If there are r possible categories then y_{ij} can be taken as an integer between 1 and r. Let

$$p_{ab}(j) = \mathrm{P}(\text{category } b \text{ at time } j + 1 | \text{category } a \text{ at time } j)$$

denote the **transition probabilities** at time j. Then the probability for the observed record $\mathbf{y}_i$ is

$$p_{y_{i1}y_{i2}}(1) \cdot p_{y_{i2}y_{i3}}(2) \cdots p_{y_{i,p-1}y_{ip}}(p-1).$$

Consequently, the likelihood function for the n individuals, i.e. for the n independent Markov chain realizations, is

$$L_n = \prod_{i=1}^{n} \prod_{j=1}^{p-1} p_{y_{ij},y_{i,j+1}}(j) = \prod_{j=1}^{p-1} \prod_{ab} p_{ab}(j)^{n_{ab}(j)}$$

where $n_{ab}(j)$ is the number of individuals who moved from category a at time j to category b at time $j+1$. Then the maximum likelihood estimator of the transition probability $p_{ab}(j)$ at time j is $\hat{p}_{ab}(j) = n_{ab}(j)/n_a(j)$, where $n_a(j) = \sum_b n_{ab}(j)$ is the number of individuals in category a at time j. In the case of stationarity $p_{ab}(j) = p_{ab}$, independent of j, and the m.l.e. becomes $\hat{p}_{ab} = n_{ab}/n_a$ where $n_{ab} = \sum_j n_{ab}(j)$ is the total number of observed transitions from a to b over the entire period and $n_a = \sum n_a(j)$ is the total occupancy of state a over the period.

A more challenging case arises when the individual profiles are not recorded, but only the **aggregate** numbers $n_a(j)$, of individuals falling into each category at each time, are available. To illustrate the problem, first consider the case of $n = 4$ individuals observed on $p = 2$ occasions with $r = 2$ categories. Suppose that there were two individuals in each category at time 1, and likewise at time 2. The likelihood for this is

$$p_{11}^2 p_{22}^2 + 4 p_{11} p_{12} p_{22} p_{21} + p_{12}^2 p_{21}^2,$$

the three terms resulting from the possibilities: (i) no movement (i.e. all four individuals stay put); (ii) partial movement (i.e. of the two individuals initially in category 1, one stays and one moves to category 2, and likewise for the other two individuals); and (iii) all individuals move. It is easy now to imagine the complex combinatorial composition of a likelihood function for larger values of n, p and r. Van der Plas (1983) gives a general, notationally neat, formula for such probabilities but confirms that a likelihood based on it would have 'unattractive' form. He therefore proposes estimation based on conditional least-squares in which the function $\sum_{j=1}^{p-1} \sum_a \{n_a(j) - m_a(j)\}^2$ is minimized. Here $m_a(j)$ denotes $E[n_a(j) \mid n_1(j-1), \ldots, n_r(j-1)]$, the conditional mean of $n_a(j)$ given the observed aggregate distribution at time $j-1$; note that $m_a(j) = \sum_b n_b(j-1) p_{ba}$. Van der Plas (1983) considers the more general case of estimating θ, a finite-dimensional parameter governing the transition probability matrix $\{p_{ab}\}$, when r may be infinite. He derives asymptotic $(p \to \infty, n$ fixed) properties for the least-squares estimator. Kalbfleisch and Lawless (1984) also note the computational intractability of the likelihood function and propose versions of conditional least-squares. Lawless and McLeish (1984) investigate the information loss due to aggregation; they compare the asymptotic variance of the least-squares estimators based on aggregate data with that of m.l.e. based on **panel** data (in which individual profiles are recorded).

Earlier work on least-squares estimators for aggregate Markov chain data includes Miller (1952) and Madansky (1959) for the case where r is finite, $\theta = \{p_{ab}\}$ and $n \to \infty$ with p fixed. A corresponding program, for continuous-time Markov processes, is carried out by Kalbfleisch, Lawless and Vollmer (1983). They use the conditional covariance matrix, with elements computed in analogous fashion to the $m_a(j)$ above, to define alternative least-squares and Gaussian estimators. Gill (1986) deals with a different kind of aggregate data in continuous time, in which only the initial distribution (the $n_a(\cdot)$ at time zero) and the 'occurrences' (the total numbers of transitions between each pair of states during the period of observation) are recorded. For estimation he uses the method of moments.

8.3 Log-linear models

An alternative approach, which is more powerful in the sense that it permits extension to higher-order cases, is via log-linear modelling. For the first-order Markov model we test for stationarity by seeing if the $(p-1)$ $r \times r$ transition probability matrices $\{p_{ab}(j)\}$ arise from the same underlying matrix of probabilities. Regarding the data as an $r \times r \times (p-1)$ table, stationarity implies that the data are generated from a log-linear model of the form

$$u + u_{1(a)} + u_{2(b)} + u_{3(j)} + u_{12(ab)} + u_{13(aj)}.$$

The $u_{12(ab)}$ corresponds to the $r \times r$ marginal, and the $u_{13(aj)}$ corresponds to the constraint that $\sum_b p_{ab}(j) = 1$ for all j.

For a second-order Markov model we have an $r \times r \times r \times (p-1)$ table. Stationarity here implies a model of the form

$$u + u_{1(a)} + u_{2(b)} + u_{3(c)} + u_{4(j)} + u_{12(ab)} + u_{13(ac)} + u_{23(bc)} + u_{14(aj)}$$
$$+ u_{24(bj)} + u_{123(abc)} + u_{124(abj)}.$$

These models are generalized in a straightforward way. In an s-order model $u_{1\ldots s}$ (and implied terms) and $u_{1\ldots s-1,j}$ (and implied term) are in the model. Thus u_{sj} and related higher-order terms are set to zero.

8.4   Linear model methods for group and time comparisons

Suppose that we have one or more groups of individuals upon whom p measures are taken. Our primary concerns are whether the groups differ (when there is more than one) and whether the times show different distributions over the responses. We will not in general be concerned with any differences in measurement profiles between individuals in the same group. It follows from this that in comparing groups or times we will be concerned with the marginal distributions over the response categories, within each group and time. That is, if $q_{gab\ldots k}$ is the probability of an individual response profile $(a, b, \ldots, k)$ in the gth group, then we will wish to base the analysis on the one-dimensional marginals $q_{ga+\cdots+}$, $q_{g+b+\cdots+}$, etc.; $+$ here indicates summation over the replaced subscript, e.g. $q_{ga+\cdots+} = P(\text{category } a$ at time 1 in group $g)$.

For comparisons between times we will test homogeneity of the p one-dimensional marginals for each group. Similarly, for comparisons between groups we will test the identity across groups of each of the p marginals. Of course, the fact that the observations are not independent across marginals for a given group means that one cannot use a straightforward approach such as simple chi-squared testing. Instead, more complex approaches are necessary.

Before exploring these it is worth remarking that tests which include patterns of differences between subjects within groups, referred to above, are derived by using the complete r^p table. The hypothesis of no difference at all between the conditions, including no differences in the way that individuals behave, is a hypothesis of total symmetry of the r^p table, i.e. $q_{ga} = q_{g\pi(a)}$ for each g and $\mathbf{a}$; here $\mathbf{a} = (a, b, \ldots, k)$ is an individual response profile and $\pi(\mathbf{a})$ is a permutation of $\mathbf{a}$. The hypothesis of no difference between groups g and h is $q_{ga} = q_{ha}$ for all $\mathbf{a}$. These more specific hypotheses together imply those above but not vice versa.

Unfortunately, log-linear model-based tests for marginal homogeneity are difficult to formulate, since the hypothesis does not have a neat description in these terms, so alternative approaches must be found.

If one wishes to use a likelihood-based approach then one can estimate the cell sizes under the model of marginal homogeneity by maximizing $L = \sum n_{ga} \log q_{ga}$ subject to $q_{gab\cdots+} = q_{gab\cdots+k} = \cdots = q_{g+b\cdots k}$. This can be done using nonlinear programming since L is convex and the constraints are linear. Once the q_{ga} have been estimated one can use the usual goodness-of-fit statistics. Unfortunately, to use this approach it is necessary to make use of all the q_{ga}, and if p is large these may be small.

An alternative approach, described by Koch *et al.* (1977), uses weighted least-squares and evades this difficulty by constructing the estimators for the one-dimensional marginals directly from the observed raw data for each subject. The dependence between marginals is taken account of by explicit use of the covariance matrix. Let (n_{ij}) be an array representing the data, where n_{ij} is the observed number in the ith population who yield the jth response profile, the latter being numbered from 1 to r^p. Let $\mathbf{n}_i = (n_{i1}, \ldots, n_{ir^p})$, $\mathbf{m}_i = \mathbf{n}_i / \sum_j n_{ij}$, and $\mathbf{m}^T = (\mathbf{m}_1^T, \ldots, \mathbf{m}_s^T)$, where s is the number of groups. A consistent estimator of the covariance matrix of $\mathbf{m}$ is then given by the $sr^p \times sr^p$ block diagonal matrix $\mathbf{V}$, with matrices

$$\mathbf{V}_i(m_i) = \{\mathbf{D}(\mathbf{m}_i) - \mathbf{m}_i\mathbf{m}_i^T\} / \sum_j n_{ij} \qquad (i = 1, \ldots, s)$$

on the main diagonal, $\mathbf{D}(\mathbf{m}_i)$ being an $r^p \times r^p$ diagonal matrix with the elements of $\mathbf{m}_i$ on its diagonal. Now let $F(\mathbf{m}) = (\mathbf{A} \otimes \mathbf{I}_s)\mathbf{m}$, with $\mathbf{A}$ a matrix producing the set of marginals for a single class and $\otimes$ the left-hand Kronecker product (the matrix $\mathbf{A}$ is multiplied by each element of $\mathbf{I}_s$). Then a consistent estimator for the covariance matrix of F is $\mathbf{V}_F = (\mathbf{A} \otimes \mathbf{I}_s)\mathbf{V}(\mathbf{A} \otimes \mathbf{I}_s)^T$. To test the hypothesis $\mathbf{CF} = \mathbf{0}$, where $\mathbf{C}$ is an appropriate contrast matrix, we can use the test statistic $(\mathbf{CF})^T[\mathbf{CV}_F\mathbf{C}^T]^{-1}\mathbf{CF}$ which, in large samples, is approximately distributed as χ_c^2, c being the number of rows of $\mathbf{C}$.

Example 8.1

To illustrate, we shall use the data presented in Bishop, Fienberg and Holland (1975, p. 308), as well as in Koch *et al.* (1977) and elsewhere, in which 46 subjects were each given three drugs. The

Table 8.1. *Response profiles for 46 patients given three drugs: F = favourable, U = unfavourable*

Response profile	Frequency
FFF	6
FFU	16
FUF	2
FUU	4
UFF	2
UFU	4
UUF	6
UUU	6
Total	46

response may be favourable (F) or unfavourable (U) to each drug. Thus $r = 2$, $p = 3$ and so there are $r^p = 8$ profiles. The data are summarized in Table 8.1.

In this case we have a single group and our interest is in whether the three drugs cause similar responses. That is, we wish to compare the marginal F/U profiles for the three drugs. The appropriate matrix **A**, with the profiles in the above order, is

$$\mathbf{A} = \begin{pmatrix} 1 & 1 & 1 & 1 & 0 & 0 & 0 & 0 \\ 1 & 1 & 0 & 0 & 1 & 1 & 0 & 0 \\ 1 & 0 & 1 & 0 & 1 & 0 & 1 & 0 \end{pmatrix};$$

Note that the constraint, that the marginals sum to 1, is satisfied. Thus $\mathbf{F} = \mathbf{AM} = (28/46, 28/46, 16/46)^T$,

$$\mathbf{V} = (46)^{-3} \begin{pmatrix} 240 & -96 & -12 & -24 & -12 & -24 & -36 & -36 \\ -96 & 480 & -32 & -64 & -32 & -64 & -96 & -96 \\ -12 & -32 & 88 & -8 & -4 & -8 & -12 & -12 \\ -24 & -64 & -8 & 168 & -8 & -16 & -24 & -24 \\ -12 & -32 & -4 & -8 & 88 & -8 & -12 & -12 \\ -24 & -64 & -8 & -16 & -8 & 168 & -24 & -24 \\ -36 & -96 & -12 & -24 & -12 & -24 & 240 & -36 \\ -36 & -96 & -12 & -24 & -12 & -24 & -36 & 240 \end{pmatrix}$$

and

$$\mathbf{V}_F = \mathbf{A}\mathbf{V}\mathbf{A}^T = (46)^{-3} \begin{pmatrix} 504 & 228 & -80 \\ 228 & 504 & -80 \\ -80 & -80 & 484 \end{pmatrix}$$

To test the identity of the three treatment profiles we can use

$$\mathbf{C} = \begin{pmatrix} 1 & -1 & 0 \\ 1 & 0 & -1 \end{pmatrix}$$

Then

$$(\mathbf{C}\mathbf{V}_F\mathbf{C}^T)^{-1} = (46^3/555312)\begin{pmatrix} 1144 & -276 \\ -276 & 552 \end{pmatrix} \quad \text{and} \quad \mathbf{C}\mathbf{F} = (0, 12/46)^T.$$

Finally,

$$(\mathbf{C}\mathbf{F})^T(\mathbf{C}\mathbf{V}_F\mathbf{C}^T)^{-1}\mathbf{C}\mathbf{F}$$

$$= (46^3/555312)(0, 12/46)\begin{pmatrix} 1144 & -276 \\ -276 & 552 \end{pmatrix}\begin{pmatrix} 0 \\ 12/46 \end{pmatrix} = 6.58$$

This is significant at the 5% level of the χ_2^2 distribution suggesting that it is improbable that the three drugs do have the same probability of producing a favourable response. ☐

The theory illustrated in the preceding example can be generalized as follows. Let $\mathbf{F}(\mathbf{m})$ be some more complex transformation of $\mathbf{m}$. Its covariance matrix will be $\mathbf{V}_F = \mathbf{H}\mathbf{V}\mathbf{H}^T$ where $\mathbf{H} = (d\mathbf{F}(\mathbf{x})/d\mathbf{x}|\mathbf{x} = \mathbf{m})$. We can then fit the model $\mathbf{F}(\boldsymbol{\mu}) = \mathbf{X}\boldsymbol{\beta}$, with $\boldsymbol{\beta}$ an unknown vector of parameters and $\mathbf{X}$ a $u \times t$ pre-specified design matrix. Here $\boldsymbol{\mu}$ is the vector of true multinomial probabilities, estimated by $\mathbf{m}$.

The test statistic for this model is $Q = (\mathbf{F} - \mathbf{X}\mathbf{B})^T\mathbf{V}_F^{-1}(\mathbf{F} - \mathbf{X}\mathbf{b})$, with $\mathbf{b} = (\mathbf{X}^T\mathbf{V}_F^{-1}\mathbf{X})^{-1}\mathbf{X}^T\mathbf{V}_F^{-1}\mathbf{F}$, and Q is approximately χ_{u-t}^2 in large samples. Given that $\mathbf{X}\boldsymbol{\beta}$ is an adequate fit to $\mathbf{F}$, one can explore restrictions on the $\boldsymbol{\beta}$ by testing $H_0: \mathbf{C}\boldsymbol{\beta} = \mathbf{0}, \mathbf{C}$ being a $c \times t$ $(c < t)$ contrast matrix, using the test statistic $(\mathbf{C}\mathbf{b})^T[\mathbf{C}(\mathbf{X}^T\mathbf{V}_F^{-1}\mathbf{X})^{-1}\mathbf{C}^T]^{-1}\mathbf{C}\mathbf{b}$. In large samples this is approximately χ_c^2 under H_0. Example 8.1 and the theory preceding it arise when $\mathbf{X} = \mathbf{I}$ and $\mathbf{H} = \mathbf{A}$.

A further common case in which there are two levels of response is that involving a logit transformation. In this case $\mathbf{F}$ is derived from $\mathbf{m}$ by a series of three transformations: $\mathbf{F}(\mathbf{m}) = \mathbf{A}_2 \log(\mathbf{A}_1\mathbf{m})$, with $\mathbf{A}_1$ producing marginals from $\mathbf{m}$ and $\mathbf{A}_2$ the logit ratios. For example, suppose that observations are made at three times, at each of which the score can be favourable (F) or unfavourable (U). Then there are eight response profiles, which we shall label m_1 to m_g in the order shown in Table 8.1. This then gives

$$\mathbf{F}(m) = \begin{pmatrix} 1 & -1 & 0 & 0 & 0 & 0 \\ 0 & 0 & 1 & -1 & 0 & 0 \\ 0 & 0 & 0 & 0 & 1 & -1 \end{pmatrix}$$

$$\cdot \log \left\{ \begin{pmatrix} 1 & 1 & 1 & 1 & 0 & 0 & 0 & 0 \\ 0 & 0 & 0 & 0 & 1 & 1 & 1 & 1 \\ 1 & 1 & 0 & 0 & 1 & 1 & 0 & 0 \\ 0 & 0 & 1 & 1 & 0 & 0 & 1 & 1 \\ 1 & 0 & 1 & 0 & 1 & 0 & 1 & 0 \\ 0 & 1 & 0 & 1 & 0 & 1 & 0 & 1 \end{pmatrix} \begin{pmatrix} m_1 \\ m_2 \\ \cdot \\ \cdot \\ \cdot \\ m_g \end{pmatrix} \right\}$$

The $\mathbf{H}$ matrix in this case becomes $\mathbf{H} = \mathbf{A}_2 \mathbf{D}(\mathbf{A}_1 \mathbf{m})^{-1} \mathbf{A}_1$. Koch *et al.* (1977) give an example of this.

Example 8.2

Example 8.1 was for a single group. In this example we compare two groups. The data for this example arose from a comparative clinical trial to investigate which of two drugs made life more tolerable for patients suffering from brain tumours. The responses analysed here are $1 =$ yes, $2 =$ sometimes, $3 =$ no, in reply to the question 'Do you still enjoy doing the things you used to?'.

We thus have two groups and three response categories, and to make the arithmetic of the example less unwieldy we shall restrict the analysis to two times, though the original data covered nine times.

The data are given in Table 8.2. Thus

$$\mathbf{n}_1^T = (5, 4, 5, 3, 4, 0, 0, 3, 1), \quad \mathbf{n}_2^T = (3, 4, 0, 5, 6, 1, 1, 2, 0)$$

and $\mathbf{m}^T = (\mathbf{n}_1^T/25, \ \mathbf{n}_2^T/22)$. The overall covariance matrix of $\mathbf{m}$ is then $\mathbf{V}(18 \times 18) = \mathrm{diag}(\mathbf{V}_1, \mathbf{V}_2)$ with, for example, $\mathbf{V}_1(9 \times 9) = 25^{-3}\{25 \mathrm{diag}(\mathbf{n}_1) - \mathbf{n}_1 \mathbf{n}_1^T\}$. The A-matrix to generate the marginals for a single day is

$$\mathbf{A}(4 \times 9) = \begin{pmatrix} 1 & 1 & 1 & 0 & 0 & 0 & 0 & 0 & 0 \\ 0 & 0 & 0 & 1 & 1 & 1 & 0 & 0 & 0 \\ 1 & 0 & 0 & 1 & 0 & 0 & 1 & 0 & 0 \\ 0 & 1 & 0 & 0 & 1 & 0 & 0 & 1 & 0 \end{pmatrix}$$

Table 8.2. *Response profiles for the two groups of patients in Example 8.2*

Profile	Frequency for group 1	Frequency for group 2
11	5	3
12	4	4
13	5	0
21	3	5
22	4	6
23	0	1
31	0	1
32	3	2
33	1	0
Total	25	22

so $\mathbf{F}$ is given by

$$\mathbf{F} = (\mathbf{A} \otimes \mathbf{I}_2)\mathbf{m} = \begin{pmatrix} \mathbf{A} & \mathbf{0} \\ \mathbf{0} & \mathbf{A} \end{pmatrix}\mathbf{m}$$

with $\mathbf{0}$ here representing a 4×9 matrix of zeros. Thus

$$\mathbf{F} = (14/25, 7/25, 8/25, 11/25, 7/22, 12/22, 9/22, 12/22)^{\mathrm{T}}.$$

A suitable contrast matrix to test between-groups differences is

$$\mathbf{C} = \begin{pmatrix} 1 & 0 & 0 & 0 & -1 & 0 & 0 & 0 \\ 0 & 1 & 0 & 0 & 0 & -1 & 0 & 0 \\ 0 & 0 & 1 & 0 & 0 & 0 & -1 & 0 \\ 0 & 0 & 0 & 1 & 0 & 0 & 0 & -1 \end{pmatrix} = (\mathbf{I}_4, -\mathbf{I}_4).$$

The test statistic Q is given by

$$Q = (\mathbf{CF})^{\mathrm{T}}(\mathbf{CV}_F\mathbf{C}^{\mathrm{T}})^{-1}\mathbf{CF} = (\mathbf{CF})^{\mathrm{T}}\{\mathbf{C}(\mathbf{A} \otimes \mathbf{I}_2)\mathbf{V}(\mathbf{A} \otimes \mathbf{I}_2)^{\mathrm{T}}\mathbf{C}^{\mathrm{T}}\}^{-1}\mathbf{CF}$$

and from the above

$$\mathbf{C}(\mathbf{A} \otimes \mathbf{I}_2)\mathbf{V}(\mathbf{A} \otimes \mathbf{I}_2)^{\mathrm{T}}\mathbf{C}^{\mathrm{T}} = \begin{pmatrix} 0\cdot0197 & -0\cdot0142 & 0\cdot0011 & -0\cdot0031 \\ -0\cdot0142 & 0\cdot0193 & 0\cdot0014 & 0\cdot0003 \\ 0\cdot0011 & 0\cdot0014 & 0\cdot0197 & -0\cdot0158 \\ -0\cdot0031 & 0\cdot0003 & -0\cdot0158 & 0\cdot0211 \end{pmatrix}$$

and $\mathbf{CF} = (0\cdot24, -0\cdot27, -0\cdot09, -0\cdot11)^T$ so that, finally, $Q = 7\cdot29$. This is not significant at the 5% level when referred to χ_4^2.

8.5 Randomization test approaches

Suppose that we have a single group containing n subjects, each of whom is measured at p times, and with the same r possible response categories at each time. We define the indicator variable x_{ija}, taking the value 1 when $y_{ij} = a$ (i.e. when the ith subject falls in category a at time j), and 0 otherwise. For the ith subject the vector $\mathbf{x}_i$ of length rp is defined by

$$\mathbf{x}_i = (x_{i11}, x_{i12}, \ldots, x_{i1r}, \ldots, x_{ipr}).$$

Test statistics to explore the distribution of the response over time can be derived based on the sum of separate components for each subject. Assuming fixed subject-by-response-category marginals $x_{i+a} (= \sum_j x_{ija})$, the null hypothesis H_0 of independence of response category and time given the subject is equivalent to random sampling of p sets of size 1 from a fixed population given by the x_{i+a}. This null hypothesis implies marginal homogeneity. Under H_0, $\mathbf{x}_i$ has conditional distribution

$$P(\mathbf{x}_i | H_0, \{x_{i+a}\}) = \prod_{a=1}^{r} x_{i+a}!/p!$$

Let $P_{ia} = x_{i+a}/p$ be the proportion of time that subject i spends in category a, and $\mathbf{P}_i^T = (P_{i1}, \ldots, P_{ir})$. Then under H_0, $E(\mathbf{x}_i | H_0, \{x_{i+a}\}) = \mathbf{m}_i = \mathbf{P}_i \otimes \mathbf{1}_p$, $\otimes$ being the left-hand Kronecker product and $\mathbf{1}_p$ a $p \times 1$ vector of 1's. Also, the conditional covariance matrix of $\mathbf{x}_i$ under H_0 is

$$\mathbf{V}_i = \{p/(p-1)\}[\mathbf{D}_i - \mathbf{P}_i\mathbf{P}_i^T] \otimes [\mathbf{I}_p - \mathbf{J}_p/p]$$

where $\mathbf{D}_i$ is the diagonal matrix with elements of $\mathbf{P}_i$ on the leading diagonal, $\mathbf{I}_p$ is the $p \times p$ identity matrix, and $\mathbf{J}_p$ is the $p \times p$ matrix of 1's.

Defining $\mathbf{A}_1$ as $(\mathbf{I}_{r-1}, \mathbf{0}_{r-1}) \otimes (\mathbf{I}_{p-1}, \mathbf{0}_{p-1})$ ($\mathbf{0}_m$ being the vector of m zeros), premultiplication of $\mathbf{x}_i$ by $\mathbf{A}_1$ will remove the last response category and last time terms from $\mathbf{x}_i$ so that $\mathbf{D}_{1i} = \mathbf{A}_1(\mathbf{x}_i - \mathbf{m}_i)$ is a vector of the differences between observed and expected frequencies under H_0 for the $u = (p-1)(r-1)$ pivot cells. The variance of $\mathbf{D}_{1i}$ under H_0 is then $\mathbf{W}_{1i} = \mathbf{A}_1\mathbf{V}_i\mathbf{A}_1^T$. Defining $\mathbf{D}_1 = \sum_{i=1}^{n}\mathbf{D}_{1i}$ and $\mathbf{V}_1 = \sum_{i=1}^{n}\mathbf{W}_{1i}$, under the null hypothesis the statistic $Q_1 = \mathbf{D}_1^T\mathbf{V}_1^{-1}\mathbf{D}_1$

follows a χ_u^2 distribution for large n and may be used to test H_0 against the general alternative that the marginal distributions of responses differ over time.

Example 8.3

To illustrate the randomization approach we shall again use the data from Example 8.1. Again we wish to test the identity of the response profiles under the three drugs.

There are 46 subjects: the first six (i.e. those having response profile FFF in Table 8.1) each have $\mathbf{x}_i = (1,0,1,0,1,0)^T$, and the next sixteen (FFU) each have $\mathbf{x}_i = (1,0,1,0,0,1)^T$, and so on. We shall illustrate the calculations for a single subject to demonstrate the procedure, and to make the exercise non-trivial we shall choose one of the FFU subjects.

For such a subject the marginal totals x_{i+a}, added over time, take values $x_{i+1} = 2$ and $x_{i+2} = 1$, so that $\mathbf{P}_i^T = (2, 1)/3$. The expected value of $\mathbf{x}_i$ is then

$$\mathbf{m}_i = \mathbf{P}_i \otimes \mathbf{I}_p = \frac{1}{3}\binom{2}{1} \otimes \begin{pmatrix} 1 \\ 1 \\ 1 \end{pmatrix} = \frac{1}{3}(2,1,2,1,2,1)^T.$$

The covariance matrix of $\mathbf{x}_i$ under H_0 is

$$\mathbf{V}_i = \frac{3}{2}\left[\frac{1}{3}\begin{pmatrix} 2 & 0 \\ 0 & 1 \end{pmatrix} - \frac{1}{9}\binom{2}{1}(2\ 1)\right] \otimes \left[\begin{pmatrix} 1 & 0 & 0 \\ 0 & 1 & 0 \\ 0 & 0 & 1 \end{pmatrix} - \frac{1}{3}\begin{pmatrix} 1 & 1 & 1 \\ 1 & 1 & 1 \\ 1 & 1 & 1 \end{pmatrix}\right]$$

$$= \frac{1}{9}\begin{pmatrix} 2 & -2 & -1 & 1 & -1 & 1 \\ -2 & 2 & 1 & -1 & 1 & -1 \\ -1 & 1 & 2 & -2 & -1 & 1 \\ 1 & -1 & -2 & 2 & 1 & -1 \\ -1 & 1 & -1 & 1 & 2 & -2 \\ 1 & -1 & 1 & -1 & -2 & 2 \end{pmatrix}.$$

For this problem

$$\mathbf{A}_1 = \begin{pmatrix} 1 & 0 & 0 & 0 & 0 & 0 \\ 0 & 0 & 1 & 0 & 0 & 0 \end{pmatrix}$$

and for this subject

$$\mathbf{D}_i = \mathbf{A}_1(\mathbf{x}_i - \mathbf{m}_i) = \begin{pmatrix} 1 & 0 & 0 & 0 & 0 & 0 \\ 0 & 0 & 1 & 0 & 0 & 0 \end{pmatrix} \left\{ \begin{pmatrix} 1 \\ 0 \\ 1 \\ 0 \\ 0 \\ 1 \end{pmatrix} - \frac{1}{3} \begin{pmatrix} 2 \\ 1 \\ 2 \\ 1 \\ 2 \\ 1 \end{pmatrix} \right\} = \frac{1}{3} \begin{pmatrix} 1 \\ 1 \end{pmatrix}.$$

Also, $\mathbf{W}_{1i} = \mathbf{A}_1 \mathbf{V}_i \mathbf{A}_i^{\mathrm{T}} = \frac{1}{9} \begin{pmatrix} 2 & -1 \\ -1 & 2 \end{pmatrix}$. Similar calculations for the other subjects lead to $\mathbf{D}_1 = \sum \mathbf{D}_{1i} = (4, 4)^{\mathrm{T}}$ and $\mathbf{u}_1 = \sum \mathbf{W}_{1i} = \frac{34}{9} \begin{pmatrix} 2 & -1 \\ -1 & 2 \end{pmatrix}$, so that $Q_1 = \mathbf{D}_1^{\mathrm{T}} \mathbf{u}_1^{-1} \mathbf{D}_1 = 8 \cdot 47$. This is significant at the 5% level of χ_2^2. $\square$

The above example tested H_0 against the general alternative hypothesis that the distributions of the responses under each of the drugs (equivalent to times) were not identical. Tests of more specific hypotheses can also be produced. For example, if the r response categories have numerical values, or can have numerical values assigned to them, then a test that the mean scores are the same for the times can be derived as follows.

Let $\mathbf{s}_i = (s_{i1}, \ldots, s_{ir})^{\mathrm{T}}$ be a vector of numerical values for the response categories for the ith subject. Let $\mathbf{A}_{2i}$ be a $p \times pr$ block-diagonal matrix with blocks $\mathbf{s}_i^{\mathrm{T}}$. Now if we let $\mathbf{D}_{2i} = \mathbf{A}_{2i}(\mathbf{x}_i - \mathbf{m}_i)$ then $\mathbf{D}_{2i}$ represents the differences between the observed and expected mean scores for the ith subject for all p times.

Parallelling the above development we now let $\mathbf{D}_2 = \sum \mathbf{D}_{2i}$ and $\mathbf{u}_2 = \sum \mathbf{W}_{2i}$ with $\mathbf{W}_{2i} = \mathbf{A}_{2i} \mathbf{V}_i \mathbf{A}_{2i}^{\mathrm{T}}$ for $i = 1, \ldots, n$. Now if $\mathbf{C}$ is a $(p-1) \times p$ matrix of $(p-1)$ contrasts spanning the space of p mean scores then $Q_2 = (\mathbf{C}\mathbf{D}_2)^{\mathrm{T}}(\mathbf{C}\mathbf{u}_2\mathbf{C}^{\mathrm{T}})^{-1}\mathbf{C}\mathbf{D}_2$ can be used to test equality of the p mean responses. Under H_0 it follows the χ_{p-1}^2 distribution for large n.

The two approaches outlined in sections 8.4 and 8.5 have different, and to a certain extent complementary, merits. The randomization method, based on individuals, permits simultaneous adjustment for all covariates. Similarly, no problems of data sparseness arise due to a possibly large number of cells in the cross-classification of the p times. A major restriction of this approach, however, is that the interchangeability hypothesis must be tested within each group of subjects (or all groups simultaneously). To apply the randomization approach to between-groups comparisons it is necessary to choose beforehand some summary function of the response profile for each subject.

In contrast, the linear model approach permits the ready comparison of groups, time changes, and group-by-time interactions, but difficulties can arise with covariates since the analysis requires that the groups be divided into subgroups by the full cross-classification of the covariates. Moreover, sample sizes need to be large enough for approximate multivariate normality to hold for the functions derived from the raw responses.

8.6 Some special cases

The case when there are just two times and two possible categories is common. One question that may be asked in such circumstances is the one treated above: are the marginals the same at the two times? This is equivalent to asking if the number who move from response 1 to response 2 between the two times is the same as the number who move from response 2 to response 1. It can be tested by a binomial test with parameter $1/2$ on the numbers who change state, to see if the observations are consistent with equal probabilities of moving each way. Equivalently, one can use the McNemar test which is, in fact, a special case of the randomization approach.

Sometimes, however, one is interested in the proportion of people who change states, rather than the absolute number. In this case a chi-squared test for independence on the table below is appropriate.

		No change	Change
	1	n_{11}	n_{12}
Response at time 1			
	2	n_{22}	n_{21}

In log-linear model terms, using the table

	Time 2	
	1	2
1	p_{11}	p_{12}
2	p_{21}	p_{22}

the saturated model is $\log p = u + u_{1(i)} + u_{2(j)} + u_{12(ij)}$ with constraints

$$\sum_i u_{1(i)} = \sum_j u_{2(j)} = \sum_i u_{12(ij)} = \sum_j u_{12(ij)} = 0.$$

The first case above tests $u_{1(i)} = u_{2(i)}$ $(i = 1, 2)$ and the second case tests $u_{2(i)} = 0$ $(i = 1, 2)$. Example 8.3 had only p conditions (or times) but only two levels of response. In such a case the randomization approach reduces to Cochran's Q test, and the algebra can be slightly simplified.

Let $x_{ij} = 1$ if the ith subject scores a success at time j, and 0 otherwise. Define $x_{i+} = \sum_{j=1}^p x_{ij}$ $(i = 1, \ldots, n)$, $x_{+j} = \sum_{i=1}^n x_{ij}$ $(j = 1, \ldots, p)$, and $\bar{x}_{++} = p^{-1} \sum_{i=1}^n \sum_{j=1}^p x_{ij}$. Then, under the null hypothesis of marginal homogeneity, the statistic

$$Q = \{ p(p-1) \sum_j (x_{+j} - \bar{x}_{++})^2 \} / \{ p \sum_i x_{i+} - \sum_i x_{i+}^2 \}$$

is approximately distributed as χ_{p-1}^2.

For the data of the example: 6 subjects have $x_{i+} = 3$, 20 subjects have $x_{i+} = 2$, 14 subjects have $x_{i+} = 1$, 6 subjects have $x_{i+} = 0$, and $x_{+1} = 28, x_{+2} = 28, x_{+3} = 16, \bar{x}_{++} = 24$. Hence $Q = 8.47$, which is significant at the 1% level using χ_2^2.

8.7 Further reading

Lee, Judge and Zellner (1977) present a comprehensive discussion of the techniques for estimating transition matrices when only aggregate data are observed. They develop and compare unrestricted, restricted, weighted and generalized least-squares estimators, minimum chi-squared estimators, maximum likelihood estimators, Bayesian methods, and minimum absolute deviation approaches. They also give computer programs, in Fortran, for calculating these estimates.

Maximum likelihood and log-linear modelling approaches are described in Bishop, Fienberg and Holland (1975). They describe the relationship between marginal homogeneity, symmetry and quasi-symmetry. Ireland et al. (1969) use a modified version of minimum discrimination information estimation and address the problem of finding estimated cell values under marginal homogeneity.

Koch et al. (1977) present a comprehensive discussion of the weighted least-squares approach to repeated measures analysis for categorical data. Landis et al. (1988) outline both the weighted least-squares linear model approach and the randomization test approach. The presentation of the latter in this chapter is based on their description. Finally, Volume 7 (1988) of *Statistics in Medicine* contains a collection of papers on repeated measures analysis for categorical data.

CHAPTER 9

Some further topics

9.1 Some practical matters

In the examples presented throughout this book 'good statistical practice' has for the most part been studiously avoided. The reason for this is simply to allow an uncluttered presentation of the current technique. Real data, as used here and unlike concocted data, almost always require more careful analysis than straightforward application of standard methodology. However, risking scorn from the experienced statistician, we have avoided getting bogged down in side issues, important though they may be in practice. We mention now briefly a few of these issues.

9.1.1 Exploratory investigation

The first thing to do with data is to look at them. There are those who claim the ability to spot odd values or relationships in a large table of figures at twenty paces. However, for humbler mortals this usually means tabulating and plotting the data in many different ways to 'see what's going on'. With the wide availability of computer packages and graphics nowadays there is no excuse for ducking the labour of this preliminary phase, and it may save some red faces later, e.g. when an important correlation is found to be just due to a misplaced decimal point in the original data. In addition to the data-screening function, the initial inspection should also suggest models to form the basis of more formal analysis. For instance, relationships between variables may be spotted, and distributions may be appraised. These visual suspicions can be augmented by small numerical investigations such as t-tests, regressions, measures of skewness, etc.

9.1.2 Transformations

The preliminary investigations described above will often suggest that the data should be transformed prior to formal analysis. It is easy to forget that most standard analyses, apart from specifically distribution-free methods, are based on a probability model for the observations. This is because to conduct such an analysis one is often not required to specify a model explicitly, e.g. when using a computer package for Anova. Nevertheless, some attention should be paid to this aspect. A commonly applied transformation for positive-valued data is to logarithms. The transformed values then have full range $(-\infty, \infty)$ and so a method based on normal distributions gains plausibility. The log-transform is also potentially useful in cases where treatment effects and errors are multiplicative. It then achieves additivity as required by linear model methods, for instance. A more sophisticated approach to transformation is to build it in as part of the model. Box and Cox (1964) introduced a family of power transformations, in which the observation y is transformed to $(y^\lambda - 1)/\lambda$, indexed by parameter λ; note that $\log y$ is obtained by continuity for $\lambda = 0$. The transformed observations are fitted by a model, usually a standard normal linear model, and λ is estimated along with all the other parameters. Many variations and ramifications have followed in the literature – one can usually find a paper on transformations in the current issues of the main statistical journals. Atkinson (1985) gives a useful survey.

9.1.3 Goodness-of-fit

After the formal analysis has been performed, and before calling the press conference to announce the findings to a rapt world, it is always a good policy to check that the model fitted fits. If you don't, others will be only too pleased to help out. First ascertain that predictions made from the model, with estimated parameters, are in line with both the data and common sense. Second, apply more technical checks. For the latter one major line of attack is via residuals. These should be calculated, examined and plotted. A source paper is Cox and Snell (1968), and the general field is covered by Cook and Weisberg (1982). At this stage one may identify outliers, i.e. observations which seem to stand apart from the rest in the context of the envisaged framework. The book by Barnett and Lewis (1978) is concerned with this subject and a recent, different approach is outlined by Chaloner and Brant (1988). Scheffé (1959, Chapter 10) gives an early account of the effect of departures from the Anova model. Also, one may identify consistent weaknesses in the model, e.g. the points follow a curve rather than the fitted straight line. For any such reason one can now take the opportunity to go round again with an improved model. Of course, the experienced practitioner will know that too much 'improvement' of the data and the model is not always productive. For instance, the current data set may be eventually fitted almost to perfection, after various modifications, only to find that next month's data do not conform at all well. A well-known instance of this is in applying stepwise regression techniques to smallish data sets.

9.1.4 Experimental design

Apart from the discussion of crossover designs in Chapter 7 very little has been said here about this undoubtedly important aspect. In our experience the statistician's first acquaintance with a study is often when the experimenter needs help to bolster up 'the stats' because his paper is under threat of rejection. It may sometimes be at an earlier stage, such as when the data are being prepared for publication. One of us vaguely recalls reading somewhere of a case where the statistician was involved before the data were collected, but this was probably in a textbook.

Of course, in an ideal world, the poor statistician, who will be lumbered with analysing the data, should have some say in its collection. On the other hand, design criteria such as efficiency (in a technical sense) are easy to apply in simple situations such as Anova with independent observations, but difficult with more involved ones, such as occur with dependent repeated measures. Thus, it may sometimes be an embarrassment to be asked for a design. Our own standard ploy is to look profound and advise the client to 'keep it simple' every time. More seriously, this policy avoids tragi-comedies such as the ruination of a beautiful Graeco-Latin square because one subject went down with 'flu half-way through, or the collapse of all the cunningly contrived confounding in a partially replicated factorial because someone dropped a test tube. Mead (1989) makes some very sensible points in this connection. By and large the statistician should make sure that the proposed study will produce data capable of answering the questions posed, and that he knows how he intends to analyse them, subject to surprises. An important point to bear in mind always is that the experimenter himself usually knows what he wants to do and, moreover, what it is feasible for him to do.

9.1.5 Multiple comparisons

Many small analyses will be performed on the data, particularly in the exploratory stage. Also, the main analysis may leave a number of subsidiary issues to be resolved, e.g. by a number of individual comparisons. In performing, say, twenty independent t-tests, at the 5% level, one of them will on average show significance when, in fact, all twenty null hypotheses are true. The position may be much worse when the tests are not independent. An example of this was discussed in section 2.1. The nub of the matter is to control the error rate so that the chance of wrongly rejecting the, possibly compound, null hypothesis is really 5%.

Some of the major statistical packages have multiple comparisons procedures as options, for instance to compare all pairs of mean values after a one-way Anova. For a general introduction to this field a good start is Miller's (1981) book. Relevant contributions here include Krishnaiah (1975), Schweder and Spjotvoll (1982), Keselman and Keselman (1984), Foutz, Jensen and Anderson (1985), Wei and Johnson (1985), Simes (1986), Hochberg and Tamhane (1987), and Hochberg (1988).

A related field, also not covered in the present account, is that of sequential testing – see Geary (1988) and the references therein.

9.1.6 Simulation

Here is the solution to those awkward theoretical problems with which people come to you with trust in their eyes. Instead of admitting that you can't calculate the expected characteristics of a standard, but complex, procedure in microbiology, say, simulate it. Most computer packages will generate pseudo-random variates on request nowadays. When done sensibly you can obtain 'exact' answers, e.g. the correct power function of a certain test, without relying on mathematical approximations such as asymptotic theory. This is particularly useful in the pharmaceutical industry where, increasingly, the regulatory authorities are asking for the powers of proposed methodology in protocols for clinical trials. Essential reading is Morgan (1984) and Ripley (1978), and Hinkley (1988) for bootstrap methodology.

9.1.7 Missing and censored observations

Most practitioners will have come to the same conclusion, that the best thing to do in this case is to fill them in while no-one's looking. We cannot recall offhand any published work on statistics which (i) contains missing or censored data, and (ii) does not specifically concern methodology for missing or censored data. To deal properly with this kind of data one must almost always sacrifice the standard, easy package option for a more involved model-fitting exercise. Censored data occur particularly frequently in survival analysis, for which we recommend Kalbfleisch and Prentice (1980), Lawless (1982) and Cox and Oakes (1984). An early paper on censored observations in ordinary straight-line regression is Glasser (1965).

9.2 Antedependence

The error covariance structures described in section 5.4.2 are capable of various generalizations. Thus, the autoregression (5.4.1) can be extended to higher orders than 1, and the error variances may be unequal, as in the second submodel of that section. Also, the time intervals between successive observations can be unequal as in section 5.4.3. A structure covering all these extensions is the so-called antedependence form of Gabriel (1961, 1962). Needless to say, the number of parameters required to formulate the extended structures is often considerably more than the former two, σ and ρ.

When the errors $e_j (j = 1, \ldots, p)$ are independent their covariance matrix $\mathbf{E}$, and hence also $\mathbf{E}^{-1}$, is diagonal. In the case of first-order Markov dependence, as in sections 5.4.2 and 5.4.3, $\mathbf{E}^{-1}$ is tridiagonal, i.e. it has one additional nonzero entry on either side of the diagonal. In the rth-order case $\mathbf{E}^{-1}$ has r nonzero entries on either side of the diagonal, i.e. it has a diagonal band of width $2r + 1$ with zeros filling the remaining top right and bottom left corners. Gabriel (1962) shows that this structure is equivalent to the following condition on the partial correlations $c_{js} = \rho(e_j, e_{j-s-1} | e_{j-1}, \ldots, e_{j-s})$: that $c_{js} = 0$ for all j and all $s > r$. In words, the partial correlation between two e's more than s epochs apart, given the intermediate e's, is zero. The extreme cases, $r = 0$ and $r = p - 1$, yield respectively complete independence (strictly speaking, zero correlation) and correlation of arbitrary pattern. In terms of autoregression, e_j is expressed as a linear function of $e_{j-1}, \ldots, e_{j-r}$ plus an independent error u_j, like equation (5.4.1). However, unlike the usual model, the regression coefficients and error variances are not necessarily constant over j.

In the framework of multivariate normal distributions correlation is synonymous with dependence, and Gabriel (1962) defines **r-antedependence** as above for this case. Suppose that a random sample of e's, each a vector like $(e_1, \ldots, e_p)$, is available. Gabriel (1962) shows that the m.l.e. under r-antedependence of $\Sigma = V(e)$ coincides with the unrestricted m.l.e., based on the usual sample covariance matrix, on the r-diagonal band and has zeros in its inverse off this band. The same property holds for Dempster's (1972) covariance selection method in which the positions of the off-diagonal zeros in Σ^{-1} are arbitary rather than being confined to the top right and bottom left corners.

Byrne and Arnold (1983) assume an antedependence structure of order 1, and consider the problem of testing for a specified μ-profile in the single-sample case.

Kenward (1987) applies the antedependence structure to some cattle growth data. There are two treatment groups, each of 30 calves, and the weights in kg were recorded at weeks $0, 2, 4, \ldots, 18, 19$, so $p = 11$ here. Kenward finds that antedependence of order 2 gives a very satisfactory fit to the sample covariance matrix, thus reducing the number of fitted covariance parameters from $p(p + 1)/2 = 66$ to $p + (p - 1) + (p - 2) = 30$. The test of fit, originally given by Gabriel (1962) as a likelihood ratio test, is shown by Kenward to be equivalent to an application of the analysis of covariance. For reasons discussed in the paper, interest centres on comparing the treatment profiles μ_A and μ_B ($p \times 1$ vectors), without explicit regression modelling, and it is shown how an overall test of $\mu_A = \mu_B$ may be broken down to examine particular sequences of times. These tests are also performed as analyses of covariance.

Kenward's procedure has the great attraction of being fairly automatic, using standard analysis of covariance methodology available in computer packages. Also, his partitioning of the test statistics yields some very useful theoretical properties, e.g. independence, though in general such advantages should not be

Table 9.1. *Pre- and post-treatment of operation data for 18 patients in two treatment groups*

No	Group	Pre	10 days	2 months	4 months	6 months	9 months	12 months
1	2	−1.00	1.55	0.84	0.99	1.01	1.07	0.57
2	1	2.11	3.22	2.11	1.95	1.76	−1.00	−1.00
3	1	−1.00	1.34	1.37	1.34	1.95	2.14	−1.00
4	2	−1.00	−1.00	−1.00	1.22	2.11	3.08	−1.00
5	1	−1.00	1.67	1.02	1.71	1.02	1.32	−1.00
6	2	−1.0	−1.00	−1.00	1.32	1.32	−1.00	1.32
7	1	−1.00	1.31	1.99	1.73	1.40	1.89	2.66
8	2	−1.00	1.00	1.36	1.42	1.10	−1.00	−1.00
9	1	−1.00	2.17	1.67	4.18	2.23	−1.00	−1.00
10	1	1.01	2.36	1.51	−1.00	4.82	−1.00	−1.00
11	2	−1.00	−1.00	2.73	1.32	−1.00	−1.00	−1.00
12	2	2.45	−1.00	1.43	1.40	1.41	−1.00	−1.00
13	2	0.93	−1.00	−1.00	3.74	2.12	1.62	1.83
14	1	2.31	2.25	−1.00	1.16	2.25	1.64	1.71
15	1	3.34	1.66	2.93	1.98	2.46	−1.00	1.63
16	2	−1.00	2.26	1.90	2.20	2.29	1.12	−1.00
17	1	−1.00	−1.00	1.21	1.95	1.62	1.25	−1.00
18	1	−1.00	−1.00	−1.00	1.16	−1.00	−1.00	−1.00
19	1	−1.00	1.98	−1.00	−1.00	1.71	−1.00	−1.00
20	2	−1.00	−1.00	1.42	1.04	−1.00	−1.00	−1.00
21	2	0.77	−1.00	1.67	1.21	1.08	1.38	−1.00
22	2	1.20	1.10	1.13	3.49	1.57	1.54	−1.00
23	1	2.70	3.50	−1.00	−1.00	−1.00	−1.00	−1.00
24	2	−1.00	−1.00	1.40	−1.00	−1.00	−1.00	−1.00
25	2	2.71	2.04	2.61	2.17	2.15	1.81	1.27
26	2	2.01	−1.00	2.03	−1.00	1.52	−1.00	−1.00
27	1	0.60	0.67	2.84	2.10	2.00	1.60	1.27
28	2	1.02	1.43	1.61	1.70	2.82	1.55	−1.00
29	2	1.71	1.71	1.21	0.90	0.61	1.66	−1.00
30	1	1.17	1.63	1.83	2.43	1.84	−1.00	−1.00
31	2	2.77	1.13	0.94	1.39	1.19	−1.00	−1.00
32	1	1.26	0.73	−1.00	−1.00	−1.00	−1.00	−1.00
33	2	1.22	1.55	1.86	−1.00	−1.00	−1.00	−1.00
34	1	1.05	1.06	1.47	1.91	2.74	−1.00	−1.00
35	1	1.51	1.38	1.44	−1.00	−1.00	−1.00	−1.00
36	2	−1.00	1.13	1.03	1.60	1.98	−1.00	−1.00
37	2	1.16	0.78	0.51	0.85	0.88	0.49	0.63
38	1	1.51	1.26	1.47	1.41	−1.00	−1.00	−1.00
39	1	−1.00	1.20	2.40	−1.00	−1.00	−1.00	−1.00
40	1	1.10	1.36	2.31	2.06	4.24	−1.00	−1.00
41	2	−1.00	2.09	1.43	1.43	1.17	1.31	−1.00
42	2	−1.00	1.00	1.44	−1.00	−1.00	−1.00	−1.00
43	2	−1.00	3.14	1.75	1.31	−1.00	−1.00	−1.00
44	1	−1.00	1.71	2.07	2.25	3.15	2.23	3.20
45	1	3.07	2.90	2.96	3.12	−1.00	3.27	−1.00
46	2	0.85	1.25	1.66	2.13	1.04	0.62	0.83
47	2	1.33	2.32	−1.00	1.72	2.04	−1.00	−1.00
48	1	−1.00	1.39	2.09	2.25	2.22	2.81	−1.00
49	1	1.42	3.40	4.10	2.92	2.65	3.40	2.25

Table 9.1. *(Contd.)*

No	Group	Pre	10 days	2 months	4 months	6 months	9 months	12 months
50	2	−1.00	0.81	1.34	1.53	1.34	0.92	0.78
51	1	−1.00	1.21	1.20	−1.00	−1.00	−1.00	−1.00
52	2	0.72	0.74	−1.00	−1.00	−1.00	−1.00	−1.00
53	1	−1.00	1.11	3.36	2.25	−1.00	−1.00	−1.00
54	2	0.60	−1.00	1.30	−1.00	−1.00	−1.00	−1.00
55	1	−1.00	2.19	1.46	3.09	−1.00	−1.00	−1.00
56	2	1.43	2.28	2.40	−1.00	−1.00	−1.00	−1.00
57	2	−1.00	1.81	2.10	2.10	2.03	1.95	−1.00
58	1	−1.00	−1.00	1.31	−1.00	−1.00	−1.00	−1.00
59	1	1.08	2.16	4.00	−1.00	−1.00	−1.00	−1.00
60	1	0.61	−1.00	0.59	0.99	1.02	0.83	0.55
61	2	−1.00	−1.00	0.74	0.90	1.00	−1.00	−1.00
62	2	−1.00	1.89	2.74	2.61	−1.00	−1.00	−1.00
63	1	−1.00	−1.00	1.82	5.30	3.09	1.42	2.31
64	2	−1.00	4.49	3.26	2.06	1.53	−1.00	−1.00
65	1	1.10	1.00	−1.00	−1.00	−1.00	−1.00	−1.00
66	2	0.60	−1.00	−1.00	−1.00	−1.00	−1.00	−1.00
67	1	1.00	1.10	5.40	2.10	−1.00	−1.00	−1.00
68	2	0.60	2.50	2.20	1.20	1.10	1.00	1.00
69	2	2.80	2.80	2.10	2.60	2.20	−1.00	−1.00
70	1	0.90	2.30	2.70	1.70	1.10	1.30	−1.00
71	2	0.90	0.80	0.70	1.00	0.80	0.60	0.70
72	1	2.60	3.60	2.60	2.90	−1.00	3.70	−1.00
73	1	−1.00	1.40	4.40	3.00	2.30	2.30	−1.00
74	2	3.40	3.30	3.40	3.40	2.10	1.50	3.10
75	2	1.10	1.20	1.50	2.40	1.50	3.20	−1.00
76	1	1.10	−1.00	3.20	3.40	3.40	2.70	2.20
77	2	4.60	1.20	3.20	2.30	2.30	1.50	−1.00
78	1	−1.00	1.90	3.90	2.20	1.40	1.60	3.40
79	2	0.60	1.00	−1.00	1.50	1.00	−1.00	1.50
80	2	1.20	1.30	−1.00	1.40	1.50	−1.00	−1.00
81	1	2.00	1.80	3.20	2.50	−1.00	−1.00	−1.00
82	1	−1.00	2.00	−1.00	−1.00	−1.00	−1.00	−1.00
83	2	1.60	0.70	1.80	2.10	1.30	1.10	1.30
84	1	1.10	1.40	1.00	2.60	0.90	2.10	1.50
85	2	−1.00	1.20	1.60	1.90	1.20	1.60	1.60
86	1	−1.00	1.40	2.90	2.90	2.40	4.10	−1.00
87	2	0.80	1.70	2.30	1.80	1.80	−1.00	−1.00
88	1	2.30	2.20	3.80	3.50	2.50	1.80	3.10
89	2	−1.00	1.58	1.56	1.35	1.96	1.33	−1.00
90	1	−1.00	2.95	0.88	2.26	−1.00	0.65	0.48
91	2	0.40	0.96	1.01	0.71	0.59	0.60	0.48
92	2	−1.00	2.05	2.03	−1.00	1.08	1.09	−1.00
93	2	1.80	1.40	1.00	1.30	2.40	2.40	−1.00
94	1	−1.00	1.60	1.40	3.40	2.60	−1.00	−1.00
95	1	2.80	2.70	−1.00	−1.00	−1.00	−1.00	−1.00
96	1	0.81	1.20	1.12	1.61	1.49	1.61	−1.00
97	1	−1.00	−1.00	3.17	2.16	3.04	2.18	−1.00
98	1	−1.00	−1.00	5.99	5.22	3.56	−1.00	−1.00

overstressed because real data have the annoying habit of not obeying the distributional rules we impose. From the modelling point of view direct application of the antedependence structure to the observations **y** rules out, for instance, random effects like those in the Anova model (section 3.1) or more generally in the two-stage model (section 5.1). For the calf growth curves one might entertain $\mathbf{y}_i = \boldsymbol{\mu}_j + \alpha_i \mathbf{1}_p + \mathbf{e}_i$, where $\boldsymbol{\mu}_i = \boldsymbol{\mu}_A$ or $\boldsymbol{\mu}_B$ and α_i is the random effect for calf i (e.g. compare calves 1 and 2 in Kenward's Table 1). With such a model $\boldsymbol{\Sigma} = V(\mathbf{y}_i) = \sigma_\alpha^2 \mathbf{J}_p + \mathbf{E}$, where $\mathbf{E} = V(\mathbf{e}_i)$, and this cannot have antedependence structure, barring artifically contrived forms for $\mathbf{E}$ involving σ_α^2.

We finish this section with another set of data for which Kenward's approach works quite nicely.

Example 9.1

Patients were each randomly assigned to receive one of two drugs. Scores were taken pre-operation, and then ten days, two months, four months, six months, nine months, and twelve months after the operation. The scores are given in Table 9.1, where -1 represents missing data.

The approach to applying the antedependence model using standard analysis of covariance routines, as outlined by Kenward (1987), involves three stages. We shall work through these in turn.

Stage 1. Finding the model order. The log-likelihood ratio for comparing a model of order r with one of order s is $-n\sum_{i=r+2}^{p} \log(e_i/d_i)$, where e_i is the residual sum of squares from an analysis of covariance with the ith measure as the dependent variable and the preceding $s_i = \min(i-1, s)$ measures as covariates, and d_i is the same thing but using only the preceding $r_i = \min(i-1, r)$ measures as covariates. Note that when $i \leqslant r+1$, $e_i = d_i$. An improved asymptotic test statistic results if this is rescaled to give

$$\Lambda = \left\{ -v \sum_{i=2}^{p} \log(e_i/d_i) \right\} - \left\{ \sum_{i=r+2}^{p} \phi(s_i - r_i, n_i - q - s_i) \right\}$$

where $v = \sum_{i=r+2}^{p}(s_i - r_i)$, n_i is the number of observations in the ith group and q is the rank of the design matrix; ϕ is given by

$$
\phi(a,m) = \begin{cases}
0 \text{ for } a = 0 \\
(2m-1)/\{2m(m-1)\} & \text{for } a = 1 \quad (m > 1) \\
2 \sum_{j=1}^{q/2} (m + 2j - 2)^{-1} & \text{for } a > 0, \quad \text{even;} \\
\phi(1,m) + \phi(q-1, m+1) & \text{for } a > 1, \text{odd.}
\end{cases}
$$

Under the null hypothesis this has an asymptotic χ_v^2 distribution.

We find an appropriate model order by sequentially comparing models of order r with those of order $s = r + 1$ until no further improvement is evident. Table 9.2 shows the derivation of Λ for the comparison with $r = 1$ and $s = 2$. Note that, since the data set has many missing values, we have only used those cases which have scores on both potential covariates (even though both were not used for $r = 1$). The 5% point for χ_5^2 is 11.07, so that adding a second covariate does not seem to improve the model. Nevertheless, the result is only just nonsignificant. We have therefore erred on the side of caution and adopted a second-order model. If one incorrectly uses too high an order the result is merely a loss of efficiency. If, however, one uses too low an order the procedure is of doubtful validity because the individual test components will not be independent.

Table 9.2. *Derivation of the statistic to compare the first and second order models*

i	r_i	d_i	s_i	e_i	$\ln(e_i/d_i)$	$n_i - q - s_i$	$\phi(s_i - r_i, n_i - q - s_i)$
3	1	31·1143	2	29·4962	−0·05341	34	0·02986
4	1	23·0238	2	21·7431	−0·05723	54	0·01869
5	1	20·8163	2	19·3411	−0·07350	53	0·01905
6	1	17·1104	2	17·1104	0·0	40	0·02532
7	1	10·8985	2	9·5810	−0·12884	19	0·05409
				Sum	−0·31298		0·14701

$$
\Lambda = \frac{(-5) \times (-0.31298)}{0.14701} = 10.64
$$

Stage 2. Do the profiles differ? Having found an appropriate order for the model, one can see if the fitted profiles for the two groups under this model differ. The test statistic is

$$-p \sum_{i=1}^{p} \log\{d_i/(d_i + h_i)\} \div \sum_{i=1}^{p} \phi(1, n_i - q - r_i)$$

where h_i is the between-groups sum of squares. Under the null hypothesis of no difference this has an asymptotic χ_p^2 distribution.

Table 9.3 gives the details of the computation of this statistic for our example. The result is significant at the 5% level, and we conclude that the profiles do differ.

Stage 3. Where do the profiles differ? Independent tests of whether the groups differ at each of the p times are given by the test statistics

Table 9.3. *Derivation of the test statistic to compare profiles*

i	d_i	h_i	$\ln(d_i) - \ln(d_i + h_i)$	$\phi(1, n_1 \cdots q \cdots r_i)$
1	43·2559	0·1542	−0·0036	0·01905
2	23·0797	1·7020	−0·0711	0·02299
3	31·1143	4·4964	−0·1418	0·02986
4	23·0238	2·3247	−0·1016	0·01869
5	20·8163	1·0290	−0·0519	0·01905
6	17·1104	0·8935	−0·0509	0·02532
7	10·8985	0·4840	−0·0493	0·05409
		Sum	−0·4702	0·18905

$$\phi \times \frac{(-7) \times (-0·4702)}{0·18905} = 17·41$$

Table 9.4. *Derivation of test statistics to explore where the profiles differ*

i	h_i	d_i	$n_i - q - s_i$	$n_i(n_i - q - s_i)$	5% level
1	0·1542	43·2559	53	0·4347	·
2	1·7020	23·0797	44	1·8013	·
3	4·4964	31·1143	34	2·2166	sig
4	2·3247	23·0238	54	2·3350	sig
5	1·0290	20·8163	53	1·6186	·
6	0·8935	17·1104	40	1·4453	·
7	0·4840	10·8985	19	0·9186	·

$\{h_i(n_i - q - s_i)/d_i\}^{1/2}$; these follow t-distributions with $(n_i - q - s_i)$ degrees of freedom under the null hypothesis.

The figures for the example are shown in Table 9.4. From this we can conclude that differences between the groups emerge at two months and four months over and above any differences attributable to correlations with earlier scores. □

9.3 Tracking

Consider for definiteness a group of growth curves, say with twice-yearly measurements of weight, in a class of schoolchildren. The concept of tracking has been described informally in various ways, such as (a) the maintenance over time of position by an individual relative to the peer group, or (b) the ability to predict individual levels from early positions within the population. The condition would occur, for instance, if the growth profiles were parallel, or just nonintersecting. Parallelism is the basis of the Anova random effects model $\mathbf{y}_i = \boldsymbol{\mu} + \alpha_i \mathbf{1}_p + \mathbf{e}_i$ (section 3.1) where $\boldsymbol{\mu}$ is the population mean profile. Where (a) has the nature of a definition of tracking, (b) focuses on its potential usefulness. Thus, sudden deviation of a child's profile from its earlier established relative position would suggest the advisability of a medical examination. In the general field of preventive health care it has been observed that extreme values in youth, of indices such as blood pressure and serum cholesterol, tend to persist into adulthood, and therefore the use of early screening is supported.

Some formal models for tracking have been proposed within the growth curve framework: $\mathbf{y}_i = \mathbf{X}_i \boldsymbol{\beta}_i + \mathbf{e}_i$ for the measurement vector $\mathbf{y}_i$ ($p \times 1$) on individual i, $\boldsymbol{\beta}_i (q \times 1) \sim N(\boldsymbol{\beta}, \mathbf{B})$ and $\mathbf{e}_i \sim N(\mathbf{O}, \sigma^2 \mathbf{I}_p)$. This is a special case of the models discussed in section 6.3. Recall that $\mathbf{y}_i \sim N(\mathbf{X}_i \boldsymbol{\beta}, \mathbf{X}_i \mathbf{B} \mathbf{X}_i^T + \sigma^2 \mathbf{I}_p)$ unconditionally and so for the jth observation time $E(y_{ij}) = \mu_j = \mathbf{x}_{ij}^T \boldsymbol{\beta}$, $V(y_{ij}) = \mathbf{x}_{ij}^T \mathbf{B} \mathbf{x}_{ij} + \sigma^2$ and $C(y_{ij}, y_{ik}) = \mathbf{x}_{ij}^T \mathbf{B} \mathbf{x}_{ik}$, where $\mathbf{x}_{ij}^T$ is the jth row of $\mathbf{X}_i$. For instance, a quadratic growth curve would be represented with $\mathbf{x}_{ij}^T = (1, \text{time}_j, \text{time}_j^2)$. Estimation and prediction of individual profiles may be performed as described in section 5.3.

McMahan's (1981) definition of tracking, for the case $\mathbf{X}_i = \mathbf{X}$ for all i, is that $E(y_{ij} | \boldsymbol{\beta}_i) = \mu_j + \kappa_i \sigma_j$ where the random effects κ_i have mean 0 and variance 1 over the population of individuals. Note that McMahan takes $\sigma_j = (\mathbf{x}_{ij}^T \mathbf{B} \mathbf{x}_{ij})^{1/2}$ (which is the same for all i) rather than the 'population standard deviation' of y_{ij} as he states. Thus κ_i measures the constant deviation of individual i from the population mean in standardized units. Note that under this model the distances apart of the individual growth profiles will vary with σ_j, so parallelism is not envisaged. McMahan defines an index of tracking $\zeta = \sigma^T X B X^T \sigma / (\sigma^T \sigma)$ where $\sigma^T = (\sigma_1, \ldots, \sigma_p)$; ζ varies from 0 to 1, the upper value reflecting perfect tracking when $X B X^T = \sigma \sigma^T$. An improved version of ζ is also given. Estimation of ζ follows straightforwardly from that of B; see section 6.2.

Foulkes and Davis (1981) discuss various indices of tracking. They mention correlations like $\rho(y_{ij}, y_{i,j+r})$ between individual measurements r time-units apart, and probabilities like $P(y_{i,j+r} > y \mid y_{ij} > y)$. The degree of tracking increases as these measures approach their upper limit 1. They propose then an index based on all p measurements $(y_{i1}, \ldots, y_{ip})$, rather than on just pairs of them, and chosen to reflect nonintersection of individual curves. The idea is that the probability of nonintersection of the profiles for any two individuals i and i' should be high, i.e. close to 1, in the case of tracking. In defining their index γ they have: (i) individuals i and i' selected at random; (ii) the tracking definition restricted explicitly to a stated time period; and (iii) profiles defined as $E(y_i \mid \text{individual } i \text{ parameters})$, i.e. with the errors e_i eliminated. Foulkes and Davis suggest estimating γ by first estimating the individual parameters, e.g. the β_i in the linear growth curve model, and then using relative frequencies of crossings of the estimated individual curves for the required probability estimate.

Ware and Wu (1981) outline estimation and prediction from growth curve analysis for longitudinal data (section 6.3.2). They are thus concerned with the aspect of tracking described in (b) in the introductory paragraph of this section, unlike the companion papers of McMahan (1981) and Foulkes and Davis (1981) which focus on (a). To venture an opinion, (b) does not seem specific enough to warrant a special term and it would seem more appropriate to reserve the particular designation 'tracking' for situations specifically like (a).

9.4 Nonlinear growth curves

The literature for this topic is less extensive than for some others covered in this book. A brief general introduction will be given here. For more discussion, applications and worked examples the following papers are representative: Fisk (1967), Crowder and Tredger (1981), Berkey (1982a and b), Crowder (1983), and Racine-Poon (1985). Sandland and McGilchrist (1979) discuss growth models based on stochastic differential equations.

9.4.1 Nonlinear regression

The models considered up till now have all been linear. Thus in $\mathbf{y} = \mathbf{X}\boldsymbol{\beta} + \mathbf{e}$ the unknown parameters, the components of $\boldsymbol{\beta}$, enter the equation linearly. A simple prototype nonlinear regression model is the exponential curve $\beta_1 + \beta_2 \exp(-\gamma x)$. Here, β_1 and β_2 enter linearly and γ is the lone nonlinear parameter; if γ were known, and only β_1 and β_2 were to be estimated, the linear model framework would suffice.

Suppose that observations y_j $(j = 1, \ldots, p)$ are recorded at settings x_j and that $y_j = \beta_1 + \beta_2 \exp(-\gamma x_j) + e_j, e_j$ being the usual zero-mean error term. This may be written in vector notation as

$$\mathbf{y} = \mathbf{X}(\gamma)\boldsymbol{\beta} + \mathbf{e} \qquad (9.4.1)$$

where $\mathbf{y}^T = (y_1, \ldots, y_p)$, $\mathbf{e}^T = (e_1, \ldots, e_p)$, $\boldsymbol{\beta}^T = (\beta_1, \beta_2)$ and $\mathbf{X}(\gamma)$ is $p \times 2$ with jth row $\{1, \exp(-\gamma x_j)\}$. This contrived way of writing the model, which goes back at least as far as Halperin (1963), has its advantages. It is now just a linear model with a small modification – the design matrix $\mathbf{X}$ depends on an unknown parameter, γ. If $\mathbf{e}$ has covariance matrix $\mathbf{E}(\phi)$, depending on parameter ϕ as in section 5.3, the normal log-likelihood for $\mathbf{y}$ is

$$L_n(\boldsymbol{\beta}, \gamma, \phi) = -\tfrac{1}{2}\{p \log(2\pi) + \log \det \mathbf{E} + Q\} \qquad (9.4.2)$$

where $Q = (\mathbf{y} - \mathbf{X}\boldsymbol{\beta})^T \mathbf{E}^{-1}(\mathbf{y} - \mathbf{X}\boldsymbol{\beta})$. Given γ and ϕ the generalized least-squares estimator obtained by minimising Q over $\boldsymbol{\beta}$, is $\hat{\boldsymbol{\beta}}(\gamma, \phi) = (\mathbf{X}^T \mathbf{E}^{-1} \mathbf{X})^{-1} \mathbf{X}^T \mathbf{E}^{-1} \mathbf{y}$; it is a function of γ through $\mathbf{X}$ and of ϕ through $\mathbf{E}$. The maximized log-likelihood function is then $L_n\{\hat{\boldsymbol{\beta}}(\gamma, \phi), \gamma, \phi\}$, a function of γ and ϕ only. As in section 5.3 this can now be maximized over γ and ϕ to find their m.l.e.'s $\hat{\gamma}$ and $\hat{\phi}$, and then $\hat{\boldsymbol{\beta}} = \hat{\boldsymbol{\beta}}(\hat{\gamma}, \hat{\phi})$ is the m.l.e. of $\boldsymbol{\beta}$.

Incidentally, just as $\mathbf{X}$ in (9.4.1) is a function of an unknown parameter, so can $\mathbf{y}$ be. Thus $\mathbf{y}$ could be replaced by a transformed version $\mathbf{y}(\lambda)$ depending on parameter λ. This is just what Box and Cox (1964) did for the linear case, with γ absent.

The form (9.4.1), together with the subsequent likelihood derivation, is generally applicable to nonlinear regression models, not just to the exponential curve used to introduce it. Other examples of nonlinear models applied in the literature include the following small but diverse collection.

1. McCullagh and Nelder (1983) refer to Hewlett's function, $\beta_1 + \beta_2 x_1 + \beta_3 x_2/(\gamma + x_2)$, in our notation. Thus $\mathbf{X}(\gamma)$ has jth row $\{1, x_{1j}, x_{2j}/(\gamma + x_{2j})\}$ here. The case $\beta_1 = \beta_2 = 0$ yields the well known (= done-to-death) Michaelis–Menten equation of pharmacokinetics.

2. Halperin (1963) discusses the regression curve $\beta_1 \exp(-\gamma_1 x) + \beta_2 \exp(-\gamma_2 x)$. This form also occurs in pharmacokinetics in modelling drug elimination under a two-compartment model (e.g. Gibaldi and Perrier, 1975). Here $\mathbf{X}(\gamma)$ has jth row $\{\exp(-\gamma_1 x_j), \exp(-\gamma_2 x_j)\}$ with $\gamma = (\gamma_1, \gamma_2)^T$ having two components.

3. Box (1971) has the example $\beta_1 \gamma_1 x_1/(1 + \gamma_1 x_1 + \gamma_2 x_2)$, which arises in describing certain chemical reaction rates, so $\mathbf{X}(\gamma)$ is $p \times 1$ with jth entry $\gamma_1 x_{1j}/(1 + \gamma_1 x_{1j} + \gamma_2 x_{2j})$ and $\gamma = (\gamma_1, \gamma_2)^T$.

4. Turner, Monroe and Lucas (1961) discuss 'rational regression' of form $(\beta_1 + \beta_2 x + \cdots + \beta_r x^r)/(1 + \gamma_1 x + \cdots + \gamma_s x^s)$, and the 'single process law', $\beta_1 + \beta_2 (x - \gamma_1)^{\gamma_2}$. Here $\mathbf{X}(\gamma)$ has jth rows $(1, x, \ldots, x^r)/(1 + \gamma_1 x + \cdots + \gamma_s x^s)$ and $\{1, (x - \gamma_1)^{\gamma_2}\}$ respectively.

5. France and Dhanova (1984) use Wood's function $\beta_1 x^{\gamma_1} e^{-\gamma_2 x}$ to describe the lactation curve in cattle. Here $\mathbf{X}(\gamma)$ is $p \times 1$ with jth entry $x^{\gamma_1} e^{-\gamma_2 x}$.

6. Mead (1970) discusses the Bleasdale–Nelder form $y = (\alpha + \beta x^\phi)^{-1/\theta}$ which has no linear parameters at all. However, to fit the relationship Mead rewrites it as either $\log y = \beta_1 \log(\alpha + \beta x^\phi)$, with $\beta_1 = -1/\theta$ and $\gamma = (\alpha, \beta, \phi)^T$ in our terms, or $y(\theta) = \alpha + \beta x^\phi$, a transformation model with $y(\theta) = y^{-\theta}$, $\beta = (\alpha, \beta)^T$ and $\gamma = (\phi)$ in our terms.

9.4.2 First model

We can now get on to repeated measurements. Suppose that the observation vector $\mathbf{y}_i (p_i \times 1)$ is recorded for individual i ($i = 1, \ldots, n$) and that a within-individuals nonlinear regression model like (9.4.1) applies. Thus

$$\mathbf{y}_i = \mathbf{X}_i(\gamma)\boldsymbol{\beta}_i + \mathbf{e}_i \qquad (9.4.3)$$

where the $\mathbf{e}_i$ are independent $N\{\mathbf{0}, \mathbf{E}_i(\phi)\}$ variates. The second stage of the model is to assume some distribution for the linear parameters $\boldsymbol{\beta}_i$ over the population of individuals. In section 6.3.2, where the nonlinear regression parameters γ were absent, the $\boldsymbol{\beta}_i$ were taken to be independently distributed as $N\{\mathbf{A}_i\boldsymbol{\beta}, \mathbf{B}(\phi)\}$, and independent of the $\mathbf{e}_i$. In consequence, the $\mathbf{y}_i$ are independent normal variates with means $\boldsymbol{\mu}_i = \mathbf{X}_i(\gamma)\mathbf{A}_i\boldsymbol{\beta}$ and covariance matrices $\boldsymbol{\Sigma}_i = \mathbf{X}_i(\gamma)\mathbf{B}(\phi)\mathbf{X}_i(\gamma)^{\mathrm{T}} + \mathbf{E}_i(\phi)$. In some applications γ and ϕ may have elements in common, so it is perhaps notationally tidier to include all the nonlinear parameters in a vector $\boldsymbol{\phi}$. Then $\boldsymbol{\mu}_i$ and $\boldsymbol{\Sigma}_i$ are just functions of $\boldsymbol{\phi}$, and $\mathbf{A}_i$ can also depend on $\boldsymbol{\phi}$ if required. The log-likelihood is then

$$L_n(\boldsymbol{\beta}, \boldsymbol{\phi}) = -\frac{1}{2} \sum_{i=1}^{n} \{ p_i \log(2\pi) + \log\det\boldsymbol{\Sigma}_i(\boldsymbol{\phi}) + Q_i(\boldsymbol{\beta}, \boldsymbol{\phi}) \} \quad (9.4.4)$$

where $Q_i(\boldsymbol{\beta}, \boldsymbol{\phi}) = (\mathbf{y}_i - \mathbf{X}_i\mathbf{A}_i\boldsymbol{\beta})^{\mathrm{T}}\boldsymbol{\Sigma}_i^{-1}(\mathbf{y}_i - \mathbf{X}_i\mathbf{A}_i\boldsymbol{\beta})$. This is essentially the same as (5.3.1) and the treatment described in that section is now applicable.

9.4.3 Second model

In (9.4.3) the nonlinear regression parameter γ does not have a subscript i, it is taken to be homogeneous over individuals, like $\boldsymbol{\phi}$ in the covariance structure. This assumption reflects an idea that γ is somehow more fundamental than the $\boldsymbol{\beta}_i$'s, i.e. more a constant of nature than the individual $\boldsymbol{\beta}$-characteristics. However, this may not be so. As a result of plotting and fitting the individual curves, which should of course always be done before embarking on the more comprehensive analysis, it may turn out that γ varies considerably between individuals. The model is now extended to accommodate this, but the previous simple structure is unfortunately mostly lost.

The within-individuals model (first stage) is

$$\mathbf{y}_i = \mathbf{X}_i(\gamma_i)\boldsymbol{\beta}_i + \mathbf{e}_i \quad (9.4.5)$$

where $\mathbf{e}_i \sim N\{\mathbf{0}, \mathbf{E}_i(\boldsymbol{\phi})\}$. For the second stage, the distribution of regression parameters over the population of individuals, suppose that $(\gamma_i, \boldsymbol{\beta}_i)$ has a joint probability density $p(\gamma_i, \boldsymbol{\beta}_i)$. This may be (multivariate) normal, but that will not ensure that $\mathbf{y}_i$ is normal because $\mathbf{y}_i$ is not linear in γ_i. The calculation of the log-likelihood is thus not so simple as before. If $p(\mathbf{y}_i | \gamma_i, \boldsymbol{\beta}_i)$ denotes the conditional (normal) density of $\mathbf{y}_i$ given the individual parameters, then the likelihood is

$$L_n(\boldsymbol{\phi}) = \log \prod_{i=1}^{n} \int p(\mathbf{y}_i|\boldsymbol{\gamma}_i, \boldsymbol{\beta}_i) p(\boldsymbol{\gamma}_i, \boldsymbol{\beta}_i) \, \mathrm{d}\boldsymbol{\gamma}_i \mathrm{d}\boldsymbol{\beta}_i; \qquad (9.4.6)$$

here $\boldsymbol{\phi}$ denotes all the unknown population parameters occurring in the distributions.

A certain amount of simplification is possible by eliminating the individual linear parameters $\boldsymbol{\beta}_i$ as follows. Let $E(\boldsymbol{\beta}_i|\boldsymbol{\gamma}_i) = \bar{\boldsymbol{\beta}}(\boldsymbol{\gamma}_i)$ and $V(\boldsymbol{\beta}_i|\boldsymbol{\gamma}_i) = \mathbf{B}(\boldsymbol{\gamma}_i)$ be the indicated conditional mean and covariance matrix of $\boldsymbol{\beta}_i$ given $\boldsymbol{\gamma}_i$. Then, from (9.4.5), $\mathbf{y}_i$ has conditional mean and covariance

$$E(\mathbf{y}_i|\boldsymbol{\gamma}_i) = \mathbf{X}_i(\boldsymbol{\gamma}_i)\bar{\boldsymbol{\beta}}(\boldsymbol{\gamma}_i), \quad V(\mathbf{y}_i|\boldsymbol{\gamma}_i) = \mathbf{X}_i(\boldsymbol{\gamma}_i)\mathbf{B}(\boldsymbol{\gamma}_i)\mathbf{X}_i(\boldsymbol{\gamma}_i)^{\mathrm{T}} + \mathbf{E}_i(\boldsymbol{\phi})$$
$$(9.4.7)$$

assuming as always that the $\mathbf{e}_i$ and $(\boldsymbol{\gamma}_i, \boldsymbol{\beta}_i)$ are independent. Note the resemblance to (6.1.2). Moreover, if the conditional distribution $\boldsymbol{\beta}_i|\boldsymbol{\gamma}_i$ is normal so is that of $\mathbf{y}_i|\boldsymbol{\gamma}_i$. Then (9.4.6) can be reduced, effectively by integrating out the $\boldsymbol{\beta}_i$, to

$$L_n(\boldsymbol{\phi}) = \log \prod_{i=1}^{n} \int p(\mathbf{y}_i|\boldsymbol{\gamma}_i) p(\boldsymbol{\gamma}_i) \mathrm{d}\boldsymbol{\gamma}_i; \qquad (9.4.8)$$

here $p(\mathbf{y}_i|\boldsymbol{\gamma}_i)$ is the multivariate normal density with mean and covariance matrix given by (9.4.7), and $p(\boldsymbol{\gamma}_i)$ is the marginal density of $\boldsymbol{\gamma}_i$ in the $(\boldsymbol{\gamma}_i, \boldsymbol{\beta}_i)$-distribution. The integrals in (9.4.8), over $\boldsymbol{\gamma}_i$ only, are of smaller dimension than those in (9.4.6).

From (9.4.7) the unconditional mean and covariance matrix of $\mathbf{y}_i$ are

$$\boldsymbol{\mu}_i = E[\mathbf{X}_i(\boldsymbol{\gamma}_i)\bar{\boldsymbol{\beta}}(\boldsymbol{\gamma}_i)]$$
$$\boldsymbol{\Sigma}_i = E[\mathbf{X}_i(\boldsymbol{\gamma}_i)\mathbf{B}(\boldsymbol{\gamma}_i)\mathbf{X}_i(\boldsymbol{\gamma}_i)^{\mathrm{T}}] + \mathbf{E}_i(\boldsymbol{\phi}) + V[\mathbf{X}_i(\boldsymbol{\gamma}_i)\bar{\boldsymbol{\beta}}(\boldsymbol{\gamma}_i)] \qquad (9.4.9)$$

The operations $E[...]$ and $V[...]$ here are taken over the $\boldsymbol{\gamma}_i$-distribution. In general $\mathbf{y}_i$ will not have a normal distribution, even though $\mathbf{y}_i|\boldsymbol{\gamma}_i$ is conditionally normal. Nevertheless, for estimation one can maximize a function with form (5.3.1) over $\boldsymbol{\phi}$. This is not the log-likelihood, but can provide useful estimates and further inference. The method is called Gaussian estimation (Whittle, 1961; Crowder, 1985).

9.4.4 Mean curves

We end this chapter with an aspect peculiar to nonlinear growth curves, where the γ_i have a nondegenerate distribution over individuals.

Example 9.2 (Crowder, 1983)

Consider the model

(i) $E(y_{ij}|\gamma_i, \boldsymbol{\beta}_i) = \beta_{i1} + \beta_{i2}\exp(-\gamma_i x_{ij})$

(ii) $\gamma_i \sim \text{gamma}\,(\xi, \eta)$, $E(\beta_{ir}|\gamma_i) = a_r + b_r\gamma_i \qquad (r = 1, 2)$

The within-individuals curve is an exponential and, across individuals, the exponential rate γ_i has a gamma distribution. In (ii) only the conditional means of β_i are specified, rather than the conditional distribution, since this will suffice. It follows that, unconditionally,

$$E(y_{ij}) = (a_1 + b_1\eta/\xi) + \{a_2 + b_2\eta/(\xi + x_{ij})\}\{\xi/(\xi + x_{ij})\}^{\eta}$$

The point to which we wish to draw attention is that $E(y_{ij})$, unlike the conditional mean, is not an exponential function of x_{ij}. $\quad\square$

In general, the conditional mean $E(\mathbf{y}_{ij}|\gamma_i, \boldsymbol{\beta}_i)$, regarded as a function of x, will define the form of the individual curves. Let us use the phrase **mean curve** to describe the curve with the same form but with the mean parameters $\{E(\boldsymbol{\gamma}), E(\boldsymbol{\beta})\}$ inserted. Thus, in Example 9.2, $E(\gamma_i) = \eta/\xi$ and $E(\beta_{ir}) = a_r + b_r\eta/\xi$, so the mean curve there has the form

$$y = (a_1 + b_1\eta/\xi) + (a_2 + b_2\eta/\xi)\exp(-\eta x/\xi).$$

On the other hand, the population mean value $E(y_{ij})$, as a function of x, has a different form, which we shall dub the **curve of means**. It defines average y-responses at given x-values, whereas the mean curve exhibits the characteristics of individual trajectories such as turning points and asymptotes. In linear models, and more generally in cases where γ_i is homogeneous over individuals as in section 9.4.2, the mean curve and the curve of means coincide because $E(y_{ij}|\boldsymbol{\gamma}, \boldsymbol{\beta}_i) = \mathbf{x}_{ij}^{\mathrm{T}}(\boldsymbol{\gamma})\boldsymbol{\beta}_i$ and $E(y_{ij}) = \mathbf{x}_{ij}^{\mathrm{T}}(\boldsymbol{\gamma})\mathbf{A}_i\boldsymbol{\beta}$, $\mathbf{x}_{ij}^{\mathrm{T}}(\boldsymbol{\gamma})$ denoting the jth row of $\mathbf{X}_i(\boldsymbol{\gamma})$.

9.5 Non-normal observations

With the exception of the treatment of binary responses in section 7.4, and categorical data in Chapter 8, all the methods covered in this book have called upon assumptions of normal distributions for the observations at some stage. This reflects traditional routine statistical practice in general for continuous measurements. In recent years, however, methodology not based on normality has come to the fore. Examples of this include the GLIM approach, based on exponential-family modelling, and survival data analysis based on the Weibull distribution. These two departures have recently begun to be developed for the repeated measures context and a brief guide is now given.

9.5.1 Generalized linear models

Wedderburn (1974) is usually made to shoulder the blame for initiating the 'quasi-likelihood' approach. Much literature, both of theory and applications, has followed, and McCullagh and Nelder (1983, 1989) are required reading.

One approach to 'maximum quasi-likelihood estimation' is via least-squares. Given independent observation vectors $\mathbf{y}_1, \ldots, \mathbf{y}_n$, with means $\boldsymbol{\mu}_1(\boldsymbol{\beta}), \ldots, \boldsymbol{\mu}_n(\boldsymbol{\beta})$ specified as functions of parameter $\boldsymbol{\beta}$, the least-squares estimator $\hat{\boldsymbol{\beta}}$ is obtained by minimizing $\sum_{i=1}^n Q_i$ where

$$Q_i = (\mathbf{y}_i - \boldsymbol{\mu}_i)^{\mathrm{T}} \boldsymbol{\Sigma}_i^{-1} (\mathbf{y}_i - \boldsymbol{\mu}_i)$$

as in section 5.2. For known $\boldsymbol{\Sigma}_i$, $\hat{\boldsymbol{\beta}}$ satisfies

$$\sum \boldsymbol{\mu}_i'^{\mathrm{T}} \boldsymbol{\Sigma}_i^{-1} (\mathbf{y}_i - \boldsymbol{\mu}_i) = 0 \tag{9.5.1}$$

obtained by differentiating $\sum Q_i$ with respect to $\boldsymbol{\beta}$; here $\boldsymbol{\mu}_i'$ is $p \times q$ with (j, k)th element $\partial \mu_{ij}/\partial \beta_k$. The linear case, $\boldsymbol{\mu}_i = \mathbf{X}_i \boldsymbol{\beta}$, yields $\boldsymbol{\mu}_i' = \mathbf{X}_i$ and thence (5.2.2).

The specialized features of generalized linear models are that: (i) $g(\mu_{ij}) = \mathbf{x}_{ij}^{\mathrm{T}} \boldsymbol{\beta}$, where g is the specified **link function** and $\mathbf{x}_{ij}$ is a vector of explanatory variables for $\mathbf{y}_{ij}$; and (ii) $\boldsymbol{\Sigma}_i = \phi^{-1} \mathbf{V}(\boldsymbol{\mu}_i)$, constraining the dependence of $\boldsymbol{\Sigma}_i$ on $\boldsymbol{\beta}$ to be through $\boldsymbol{\mu}_i$ only, via the specified **variance function** $\mathbf{V}$ (a matrix of specified functions) and scaling parameter ϕ (possibly unknown). The equation (9.5.1) is now used to define, as its solution, the maximum quasi-likelihood estimator $\tilde{\boldsymbol{\beta}}$. Note that this is no longer equivalent to the minimization of $\sum Q_i$

since differentiation would now introduce extra terms involving $\partial \mathbf{\Sigma}_i / \partial \beta_j$. However, as it stands, (9.5.1) is an unbiased estimating equation usually yielding well-behaved solutions (McCullagh, 1983), though not always (Crowder, 1987). In addition, for the univariate case ($p = 1$), (9.5.1) can often be identified with the likelihood equation corresponding to an exponential-family distribution for the observations. This is often taken to be the primary derivation for (9.5.1).

Liang and Zeger (1986) and Zeger and Liang (1986) propose partial specification of the model. They use a link function g as above, and a variance function $\mathbf{V}(\cdot)$ such that $\sigma_{ij}^2 = (\mathbf{\Sigma}_i)_{jj} = \phi^{-1} \mathbf{V}(\mu_{ij})$, but do not assume precise knowledge of the covariances $(\mathbf{\Sigma}_i)_{jj'}$. Thus, only the marginal moments of the y_{ij}'s, the means and variances, are explicitly modelled. Then, to use (9.5.1) for estimation of $\boldsymbol{\beta}$, they replace $\mathbf{\Sigma}_i$ there by $\phi^{-1} \mathbf{A}_i^{1/2} \mathbf{R}_i(\boldsymbol{\alpha}) \mathbf{A}_i^{1/2}$, where $\mathbf{A}_i = \text{diag} \{\mathbf{V}(\mu_{ij})\}$ and $\mathbf{R}_i(\boldsymbol{\alpha})$ is a 'working' correlation matrix depending on a (possibly unknown) parameter vector $\boldsymbol{\alpha}$. If $\mathbf{R}_i(\boldsymbol{\alpha})$ is taken as $\mathbf{I}_p$, the $p \times p$ unit matrix, (9.5.1) reduces to the 'independence equations', i.e. the set of equations which would be 'correct' if $\mathbf{\Sigma}_i$ were truly diagonal. Other possible choices for $\mathbf{R}_i(\boldsymbol{\alpha})$ are suggested in Liang and Zeger (1986), including the structures mentioned above in sections 5.4.1 and 5.4.3. If $\boldsymbol{\alpha}$ is known, (9.5.1) is sufficient alone for estimation of $\boldsymbol{\beta}$. Otherwise $\boldsymbol{\alpha}$ must itself be estimated and the papers propose see-saw methods for this. Some asymptotic properties are also given, together with some efficiency calculations. The second paper works through a numerical example.

Zeger, Liang and Albert (1988) discuss SS (subject specific) and PA (population average) models for longitudinal data. Explicitly, the PA model is as described in the 1986 work reviewed above, with link $g_{\text{PA}}(\mu_{ij}) = \mathbf{x}_{ij}^{\text{T}} \boldsymbol{\beta}$ and variance $\sigma_{ij}^2 = \phi^{-1} \mathbf{V}_{\text{PA}}(\mu_{ij})$. On the other hand, the SS model has $g_{\text{SS}}(m_{ij}) = \mathbf{x}_{ij}^{\text{T}} \boldsymbol{\beta} + \mathbf{z}_{ij}^{\text{T}} \mathbf{b}_i$ and $s_{ij}^2 = \phi^{-1} \mathbf{V}_{\text{SS}}(m_{ij})$, where $m_{ij} = E(y_{ij}|\mathbf{b}_i)$ and $s_{ij}^2 = V(y_{ij}|\mathbf{b}_i)$ are the indicated conditional moments; the $\mathbf{b}_i$ are random coefficients with some distribution over the population of individuals. The analogy with the two-stage model of section 6.1 is obvious. Zeger et al. (1988) discuss the relative merits, and the differing interpretations and consequences, of the two types of model. For estimation, the 1986 work is used as before for the PA models. For the SS models (9.5.1) is applied, for which $\boldsymbol{\mu}_i$ and $\mathbf{\Sigma}_i$ need to be evaluated from m_{ij}, s_{ij}^2 and the assumed $\mathbf{b}_i$-distribution. If the latter contains unknown parameters an extra step in the estimation cycle is necessary. The paper refers to several cases of the models which have appeared previously in the literature (e.g. Stiratelli, Laird and Ware, 1984), and works through an example.

Moulton and Zeger (1989) use standard maximum quasi-likelihood estimation (m.q.l.e.), based on the univariate version of (9.5.1), to estimate β from the data at each separate time point, thus obtaining $\tilde{\beta}_1,...,\tilde{\beta}_p$. This enables examination of possible trends in β over time; in the absence of such trend an overall estimate may be computed by averaging the $\tilde{\beta}_j$s in some way. The distributional modelling, upon which the estimation is based, is thus marginal, i.e. for separate components of y_i as in Liang and Zeger (1986) and Zeger and Liang (1986), rather than multivariate. The bootstrap is proposed for inference about parametric functions $\psi(\beta_1,...,\beta_p)$. A numerical example is given.

Zeger and Qaqish (1988) use m.q.l.e. for time series with covariates. The material has some relevance here and we give a brief review slanted towards the present section. Let $m_{ij} = E(y_{ij}|D_{ij})$ and $v_{ij} = V(y_{ij}|D_{ij})$ be the conditional mean and variance given D_{ij}, the outcome and covariate history for individual i prior to time j. The model comprises $g(m_{ij}) = x_{ij}^T\beta + f_{ij}^T\theta$ and $v_{ij} = \phi V(m_{ij})$ as in the standard GLM specification above except for the term involving f_{ij}, a vector of given functions of D_{ij}; β and θ are the unknown regression parameters. If the f_{ij} capture only a limited number of past values $(y_{i,j-1},...,y_{i,j-l})$, say, for some fixed lag l), and if assumptions are made to ensure that the conditional distribution of y_{ij} given D_{ij} is defined by m_{ij} and v_{ij}, then the y-process is Markov (of order l). The examples cited in the paper are of this type. The authors go on to describe estimation, based on (9.5.1), for the case $n = 1$ and with a canonical link function, and analyse some data. Zeger (1988b) makes another contribution to this area, but with a different model formulation.

9.5.2 Mixed Weibull models

The Weibull distribution, with regression modelling, is widely used for the parametric analysis of survival and reliability data. Standard treatments include Kalbfleisch and Prentice (1980), Lawless (1982), and Cox and Oakes (1984). These works deal in the main with independent observations.

For repeated measures Crowder (1985, 1989) has suggested the following approach. Suppose that $y_1,\ldots,y_p$ are the measures in question, supposedly Weibull-distributed and independently made on the individual. If y_j has the Weibull survivor function $\exp(-\xi_j t_j^{\phi_j})$ then $\log y_j$ has mean and variance given by

$$E(\log y_j) = -\phi_j^{-1}(\log \xi_j + \gamma), \quad V(\log y_j) = \pi^2/(6\phi_j^2)$$

where $\gamma = 0\!\cdot\!5772$ (Euler's constant). By analogy with the normal random effects model it is assumed that ϕ_j is homogeneous over individuals and that $\log \xi_j = \mu_j + \alpha_i$, where μ_j represents the fixed effects (modelled, e.g. linearly as $\mathbf{x}_{ij}^T\boldsymbol{\beta}$) and α_i is the random individual level. In consequence, the joint survivor function of $\mathbf{y} = (y_1,\ldots,y_p)$ is given by

$$\bar{F}(y) = \int \exp(-se^{\alpha}) \, dG(\alpha)$$

where $s = \sum_{j=1}^{p} e^{\mu_j} t_j^{\phi_j}$ and $G(\alpha)$ is the distribution function of α (over the population of individuals). Tractable forms for $G(\alpha)$, i.e. which yield a usable form for $\bar{F}(y)$, are given by Crowder (1985, with a gamma distribution for α) and Crowder (1989, a positive stable distribution). In each paper, the resulting joint distributions are examined and applied to data from response-time studies. Missing and right-censored observations are easily accommodated.

CHAPTER 10

Computer software and examples

10.1 Repeated measures facilities in BMDP, SPSS^x and SAS

There are many software packages which will perform various kinds of repeated measures analyses. Some of them have special routines which produce output in a form specially suited to repeated measures. Others permit repeated measures analyses to be performed as special cases of more general models. And yet others have no special facilities but permit macros to be written which will perform repeated measures analyses. Obviously it is impracticable to write a comprehensive review of all that is available, including code for macros and so on. Moreover, the problem is compounded by the constant development of software, both new and old. New programs are introduced and old ones are updated as inadequacies are identified or new developments incorporated. Despite these complications, however, we believe it would be useful to describe briefly the facilities which are available in recent issues of some of the most comprehensive and widespread computer packages. Thus, in what follows we summarize the repeated measures analysis facilities available in BMDP, SPSS^x and SAS. Only a summary is given since it would be inappropriate here to reproduce the full details and syntax of the particular programs involved. Those can be found in the relevant manuals. Here we have just attempted to single out the most important aspects.

Of the programs outlined below, BMDP 5V, SAS GLM and SPSS^x MANOVA in particular are very flexible. They relate to the 1988 software version of BMDP, version 5 of SAS, and release 2·1 of SPSS^x. Further information can be obtained from:

BMDP Statistical Software Inc.
1440 Sepulveda Boulevard,
Suite 316,
Los Angeles, CA 90025, USA.
(BMDP is a registered trademark of BMDP Statistical Software Inc.)

SPSS Inc.,
444 N. Michigan Avenue,
Chicago,
Illinois, 60611,
USA.
(SPSS^x is a registered trademark of SPSS Inc.)

SAS Institute Inc.,
SAS Circle,
Box 8000,
Cary, NC 27511-8000,
USA.
(SAS is a registered trademark of SAS Institute Inc.)

BMDP 2V

This program performs univariate fixed effects analysis of variance and covariance and repeated measures analyses for designs involving several groups, which can be formed from crossed but not nested factors, and for crossed within-subjects factors. Unequal group sizes are allowed, but all observations must be present for each subject measured. The hypotheses tested are unweighted sums of cell means. That is, each effect is tested by comparing models with and without the effect (Yates's weighted squares of means analysis).

The terms to be included in the between-groups model can be specified. The raw repeated measurements are transformed to orthogonal polynomials (other orthogonal contrasts are not available) and the user can request tests on individual components to be presented. Unequally spaced times or levels can be specified.

The program gives unadjusted p-values for the within-subjects factor effects and their interactions and gives the results of a sphericity test so that the user can judge whether an adjustment is needed. Greenhouse–Geisser and Huynh–Feldt epsilons and their associated adjusted p-values are given.

Covariates may be constant or vary across time, and individual cases can be given weights to reflect the accuracy with which measurements have been taken on each case.

BMDP 3V

This uses maximum likelihood and restricted maximum likelihood methods to analyse fixed and random effects analysis of variance and covariance. All cases must have complete data.

BMDP 4V

This program performs univariate and multivariate analysis of variance, covariance and repeated measures. It permits the user to specify the weights to be used with each cell in an analysis. Thus one can use equal weights to analyse the unweighted means hypotheses (Yates's weighted squares of means analysis) as in BMDP 2V or one can set weights equal to cell sizes (which is equivalent to testing each effect with others constrained to zero) or one can choose other sets of weights. Since the equal weights analysis includes interactions in the model when main effects are tested, which may not be regarded as appropriate, this program has the facility to specify that the interaction should not be included in the model.

Multivariate statistics output are Wilks's lambda, the Hotelling–Lawley trace, Hotelling's generalized T-zero squared statistic, and Roy's largest root statistic. Univariate test results are also presented with Greenhouse–Geisser and Huynh–Feldt adjustments, along with the associated epsilons.

Full explanations of some of the commands used by 4V, including the facility for nested repeated measures factors, are not given in the manual but in separately available technical reports.

BMDP 5V

This program uses maximum likelihood or restricted maximum likelihood methods to tackle repeated measures problems with general covariance matrices. Covariance matrix forms permitted include compound symmetry, autoregressive, completely general, and some requiring structures to be specified in terms of parameters (e.g. factor analytic, random coefficients). There is also a facility for specifying covariance structures which are not linear and are not of the special cases specified, by means of user-supplied FORTRAN subroutines defining the covariance matrix.

For the unrestricted maximum likelihood estimates, this program has available Newton–Raphson, Fisher scoring and generalized EM algorithms, and for the restricted maximum likelihood estimates it has a generalized EM algorithm and a quasi-scoring algorithm.

For each term in the model, 5V gives a test result of the null hypothesis that the term can be omitted from the model. By running the program to fit several models one can construct likelihood ratio tests of hypotheses that the covariance matrix takes particular forms.

There is some capacity for handling missing values, and the program permits time-varying covariates.

BMDP 8V

This is a program for general mixed model analysis of variance for any complete design (nested, crossed, partially crossed, random or fixed effects). It does not handle unequal cell sizes or covariates and does not perform orthogonal decompositions, permit contrasts to be specified, carry out range tests, or do multivariate analysis of variance.

SAS ANOVA

This is a general univariate and multivariate analysis-of-variance program for balanced data. For repeated measures the program gives both univariate and multivariate tests and tests on single degree of freedom contrasts. Contrasts permitted on the repeated factors are: one level versus the others, polynomials, Helmert contrasts, each level versus the mean of the others, and successive pairwise differences.

Sphericity tests to explore the legitimacy of the univariate analysis can be conducted. Multivariate statistics calculated are Wilks's lambda, the Pillai trace, the Hotelling–Lawley trace, and Roy's maximum root criterion.

SAS GLM

GLM fits general linear models using least squares. It can handle unbalanced data and its scope includes repeated measures analysis of variance and multivariate analysis of variance. Contrasts available on the repeated factors include one level versus each of the others, polynomials, Helmert contrasts, each level versus the mean of the others, and successive pairwise differences. Tests on single degree of freedom contrasts can be conducted. Spacings of the levels can be specified for the polynomial contrasts.

The univariate test results are also presented with Greenhouse–Geisser and Huynh–Feldt adjustments and their associated epsilons. Sphericity tests can be requested, and multivariate statistics output are Wilks's lambda, the Pillai trace, the Hotelling–Lawley trace, and Roy's maximum root criterion.

Four approaches to calculating sums of squares are used: adjusting each effect for preceding effects in a user-specified order, adjusting each effect for all other effects, and two methods which coincide with unweighted means hypotheses (Yates's weighted squares of means analysis) if there are no empty cells in factorial models.

SAS CATMOD

There seem to be very few generally available programs specially suited to repeated measures analysis for categorical data. CATMOD is one of these few. CATMOD fits linear models to categorical data using weighted least squares. It is based on the approach outlined in Chapter 8 where it is assumed that the primary interest is a comparison of the marginal distributions over the response categories within each cell of the group by time cross-classification.

Since the tests are based on asymptotic theory, one needs to be confident that the sample sizes are adequate for the approximations to be valid. In the manual (*SAS Users Guide: Statistics*, Version 5 edition, p. 205) it is suggested that an effective sample size of 25–30 cases is needed for each response function being analysed.

Cases with missing values are omitted from the analysis.

SPSS^x MANOVA

This is a general univariate and multivariate analysis of variance and covariance program for any crossed or nested design. It can handle unbalanced data and offers two methods of partitioning the sums of squares: each term corrected for every other term in the model (the default) and hierarchical decomposition in which each term is adjusted only for those terms which precede it.

Contrasts available on the levels of the repeated factors include: one level versus each of the others, polynomials, Helmert contrasts, each level versus the mean of the others, successive pairwise differences, and user-specified sets of contrasts. The program will present single degree of freedom test results.

Sphericity tests can be conducted and multivariate statistics presented are Wilks's lambda, the Pillai trace, the Hotelling–Lawley trace, and Roy's maximum root criterion.

The program also has extensive internal diagnostic plotting facilities.

SPSS^x RELIABILITY

This program performs an item analysis by computing commonly used coefficients of reliability but can also be used to perform basic univariate repeated measures analyses.

For ranked data Friedman's chi-squared statistic can be obtained.

Just as we cannot describe all the available repeated measures software, so we cannot give examples of all of it. However, we have found it useful for students to have annotated examples of the sorts of things one might expect from modern repeated measures analysis programs. For this reason, we include three annotated sample listings from BMDP 2V and two from SPSS^x MANOVA in the following sections.

10.2   Example 1 – BMDP program 2V

Program 1:

BMDP program P2V for analysis of variance and covariance, including
repeated measures.

The data analysed here are described in detail in example H of Hand and
Taylor (1987) and are shown in Table 10.1. Subjects are cross-classified
by two factors: COND and PREF. Twelve measurements are taken on each
subject under different conditions. These twelve measurements are formed
by a cross-classification of two factors: trial (1 to 4) and time within
trial (1 to 3; called 'epoch'). The variables are measured in the order
T1E1, T2E1, T3E1, etc. In this analysis the factorial structure of the
variables is ignored.

```
[1] /PROBLEM   TITLE='HAND AND TAYLOR (1987) P.194'.
[2] /INPUT   VARIABLES=14.   UNIT=10.   FORMAT IS '(1X,2F1.0,12F5.0)'.
[3] /VARIABLE NAMES ARE COND,PREF.
[4] /TRANSFORM  X(3)=LN(X(3)+40).
                X(4)=LN(X(4)+40).
                X(5)=LN(X(5)+40).
                X(6)=LN(X(6)+40).
                X(7)=LN(X(7)+40).
                X(8)=LN(X(8)+40).
                X(9)=LN(X(9)+40).
                X(10)=LN(X(10)+40).
                X(11)=LN(X(11)+40).
                X(12)=LN(X(12)+40).
                X(13)=LN(X(13)+40).
                X(14)=LN(X(14)+40).
[5] /DESIGN   DEPENDENT ARE 3 TO 14.
[6]           LEVEL IS 12.
[7]           GROUPING IS COND,PREF.
    /PRINT
    /END
```

----

[1] Gives a title for the program.
[2] Describes how many variables are to be read, and their format.
[3] Names the first two variables (which are in fact the grouping
    factors).
[4] Transformations to correct for skew distributions.
[5] The DESIGN paragraph specifies the factors and dependent variables.
    Variables 3 to 14 are the dependent variables.
[6] There are twelve dependent variables (the response is measured at
    twelve times).
[7] COND and PREF are the two grouping factors.

Table 10.1. *Data for Example 1. The first two columns are between–subjects grouping variables: condition and preference, each with two levels. The other 12 variables are heart rate (× 100) measured at three times (epochs) within each of four trials. The order of observations is (time1, epoch1), (time2, epoch 1), (time 3, epoch 1), etc.*

6114	6794	6663	6644	6381	6817	6433	6522	6245	6873	6372	6330
7178	6878	7431	7216	6995	6967	7211	7042	6972	7154	7206	7009
6902	6630	6311	6545	6883	6442	6433	6897	6967	7141	6353	6916
6573	6348	6414	6381	6780	6752	6963	6658	6817	6831	7202	6770
6606	7075	6438	6508	6522	6250	6344	6452	6447	6620	6405	6386
8125	7938	7652	7548	7970	8177	8477	7323	7970	8439	7708	7389
6381	6044	6133	6902	6484	6255	6213	6644	7230	6508	5866	6611
6813	7248	6658	7042	6892	6882	6550	6592	6925	6677	6302	6775
7066	6939	7206	7113	7136	6789	7047	7239	6977	6883	7174	7094
10009	10539	9030	9925	9691	10727	9803	9714	10338	10694	9869	9911
9409	8467	9302	9255	9302	8964	8308	9264	9639	8734	9409	8598
7197	7094	7309	7169	7394	7267	7230	7197	7436	7811	7263	7089
6728	6822	6981	7042	6752	6728	7108	6906	6700	6709	6536	6948
6841	6105	6588	6564	7323	6620	6606	6536	7225	6203	6020	7103
5734	5861	5758	5477	5631	6030	5828	6606	6081	6353	6095	6995
6489	5913	5809	5903	7145	5880	6063	5913	8139	6306	6555	6405
6405	6775	6948	6227	7455	6850	7506	6630	7009	6841	7202	6311
6573	6948	6630	6592	6995	6775	6742	7061	6878	6756	6808	6086
7333	7323	7877	7539	7464	7281	7698	7792	7258	7356	7277	7778
6189	5982	5861	5988	6020	6044	6072	5730	6306	6245	5903	6152
5336	5720	5223	5013	5303	5383	5265	5298	4952	5477	5238	5125
6302	6142	6072	6227	6465	6217	6386	6264	6808	5875	6142	6152
7558	7009	7216	6123	7769	7427	6930	6602	6536	6456	7281	7014
6367	6761	6452	7155	6494	6213	6770	6414	7094	6958	6784	6663
5898	5763	5898	5655	5847	5777	6123	5688	5894	6025	5767	5748
6166	5898	5828	5800	5955	5866	5870	5973	6170	5791	5875	6100
5488	5173	4866	6057	4727	4823	5102	5788	6001	5204	4879	4915
6048	5800	5645	5856	6250	5748	5936	5908	6367	5884	5828	5772
8252	8069	8514	8575	8266	8125	7825	8430	8336	8275	8341	7933
8336	8702	8523	8073	8284	8584	8317	8041	8331	8463	8185	8144
6223	6082	6404	6047	6316	5982	7020	6035	6322	6129	5695	6052
6128	6132	6207	6306	6063	6180	6189	6419	6189	6292	6264	6460
7773	7005	8205	7525	7773	7234	7788	7543	8031	7530	7525	6766
7220	7033	7459	7098	7028	6798	7380	7159	7089	6939	7323	7113
5495	5486	5603	5552	5664	5481	5336	5523	6189	5378	5341	5598
6573	6278	5420	6175	6958	6625	5383	5730	6859	6386	5561	6030
6269	6400	6963	7244	5809	6072	6634	6625	6522	6030	6625	6761
6423	6611	6531	6536	6283	6536	6283	6906	6395	6588	6344	6545
7863	7080	7455	7267	7305	7103	7248	7319	7272	7183	7155	7192
6756	6372	6470	5711	6222	6325	6705	6152	6170	5941	6339	5997
8406	8772	8922	7938	7825	7695	8584	7713	7755	7675	8097	8177
7080	6808	6841	7052	6963	7267	6700	6564	6930	6878	6939	6409
7056	6231	6358	6461	6836	6311	6400	6452	7267	6255	6611	6686
6827	6789	6728	6663	6653	6522	6723	6878	6719	6583	6569	6822
8275	7014	7778	7933	8598	7534	7445	7951	8397	7380	7389	7708
10131	9203	9292	9048	10300	9630	9044	9048	10853	9855	9034	9091
7234	7173	8102	7225	7591	6911	7614	7492	7919	6939	7741	7741
6072	5913	6259	6278	5889	6363	5856	5945	5917	6119	5908	5997
6419	6363	7164	7567	6714	6583	6803	7038	5992	6470	6700	6611
6916	7867	7469	6944	6423	6761	6789	7173	6564	7225	6583	7005
6644	6784	6817	6845	6934	6752	6770	6892	6859	6652	6920	6752
5608	5613	5748	5832	5683	5894	5772	5913	6175	6250	5706	5861
7286	7197	7127	7952	7347	7244	7192	6672	7239	7413	7084	7647
8613	9297	8570	9058	8777	8533	8519	8913	8870	8805	8603	9339
6986	6911	7811	7525	6738	6766	7188	7080	6550	6733	6944	6672
6109	6414	5866	5856	6105	6461	5950	6006	5856	5969	5884	5589
5636	5458	5641	5988	5575	5472	5575	5894	6067	5781	5491	5716
6189	6034	5781	5941	5777	6072	5800	5734	6077	5931	5875	5734
5369	5659	5327	5636	5514	5682	5505	5655	5584	5650	5486	5603
5776	6020	6170	5898	5523	5608	6494	5645	5702	5561	6320	5420

These data are described in detail in Example H of Hand and Taylor (1987).

Output from program 1:

PROBLEM TITLE IS
HAND AND TAYLOR (1987) P.194


```
NUMBER OF VARIABLES TO READ IN. . . . . . . .          14
NUMBER OF VARIABLES ADDED BY TRANSFORMATIONS. .          0
TOTAL NUMBER OF VARIABLES . . . . . . . . . . .          14
NUMBER OF CASES TO READ IN. . . . . . . . . . . TO END
CASE LABELING VARIABLES . . . . . . . . . . .
MISSING VALUES CHECKED BEFORE OR AFTER TRANS. . NEITHER
BLANKS ARE. . . . . . . . . . . . . . . . . . . MISSING
NPUT UNIT NUMBER . . . . . . . . . . . . . . .          10
REWIND INPUT UNIT PRIOR TO READING. . DATA. . .      YES
NUMBER OF WORDS OF DYNAMIC STORAGE. . . . . . .     351230
NUMBER OF CASES DESCRIBED BY INPUT FORMAT . . .           1
```

VARIABLES TO BE USED

1 COND	2 PREF	3 X(3)	4 X(4)	5 X(5)
6 X(6)	7 X(7)	8 X(8)	9 X(9)	10 X(10)
11 X(11)	12 X(12)	13 X(13)	14 X(14)	

INPUT FORMAT IS
(1X,2F1.0,12F5.0)

MAXIMUM LENGTH DATA RECORD IS   63 CHARACTERS.

I N P U T   V A R I A B L E S . . . . .

VARIABLE NO.	NAME	RECORD NO.	COLUMN BEG	END	INPUT FORMAT	VARIABLE NO.	NAME	RECORD NO.	COLUMN BEG	END	INPUT FORMAT
1	COND	1	2	2	F1.0	8	X(8)	1	29	33	F5.0
2	PREF	1	3	3	F1.0	9	X(9)	1	34	38	F5.0
3	X(3)	1	4	8	F5.0	10	X(10)	1	39	43	F5.0
4	X(4)	1	9	13	F5.0	11	X(11)	1	44	48	F5.0
5	X(5)	1	14	18	F5.0	12	X(12)	1	49	53	F5.0
6	X(6)	1	19	23	F5.0	13	X(13)	1	54	58	F5.0
7	X(7)	1	24	28	F5.0	14	X(14)	1	59	63	F5.0

DESIGN SPECIFICATIONS

```
          GROUP  =    1   2
          DEPEND =    3   4   5   6   7   8   9  10  11  12  13  14
          LEVEL  =   12
```

BASED ON INPUT FORMAT SUPPLIED   1 RECORDS READ PER CASE.

| VARIABLE | | STATED VALUES FOR | | | GROUP | CATEGORY | INTERVALS | |
NO.	NAME	MINIMUM	MAXIMUM	MISSING	CODE	INDEX	NAME	.GT.	.LE.
1	COND				1.000	1	*1		
					2.000	2	*2		
2	PREF				1.000	1	*1		
					2.000	2	*2		

NOTE: CATEGORY NAMES BEGINNING WITH * WERE CREATED BY THE PROGRAM.

NUMBER OF CASES READ.   60
PAGE   2 BMDP2V HAND AND TAYLOR (1987) P.194

GROUP STRUCTURE

COND	PREF	COUNT
*1	*1	20
*1	*2	10
*2	*1	17
*2	*2	13

SUMS OF SQUARES AND CORRELATION MATRIX OF THE
ORTHOGONAL COMPONENTS POOLED FOR ERROR 2 IN ANOVA TABLE BELOW
[1]

```
        0.11761     1.000
        0.06928    -0.030   1.000
        0.09531    -0.049  -0.161   1.000
        0.07361    -0.091   0.128   0.043   1.000
        0.09413     0.054  -0.314   0.230   0.187   1.000
        0.21110    -0.308  -0.190   0.135   0.211   0.133   1.000
        0.20985     0.051  -0.045  -0.414  -0.301  -0.257  -0.070   1.000
        0.09907     0.206   0.046  -0.316  -0.185  -0.139  -0.587   0.179
        0.11801    -0.036   0.010   0.404   0.455   0.485   0.146  -0.365
        0.06006     0.048   0.127  -0.161   0.215  -0.058  -0.269   0.043
        0.12728     0.037  -0.107   0.161   0.180   0.217  -0.079   0.114

        0.09907     1.000
        0.11801    -0.383   1.000
        0.06006     0.230   0.123   1.000
        0.12728    -0.016   0.515  -0.004   1.000
```
[2]
SPHERICITY TEST APPLIED TO ORTHOGONAL COMPONENTS - TAIL PROBABILITY   0.0000
PAGE   3 BMDP2V HAND AND TAYLOR (1987) P.194

----

[1] The within groups correlation matrix of orthogonal contrasts of the
variables should be diagonal with identical elements on the diagonal for a
valid univariate analysis, unless some adjustment is made, as explained in
Chapter 3. This is the matrix.

[2] The results of a test of sphericity show that in this case the
condition seems not to be true.

CELL MEANS FOR   1-ST DEPENDENT VARIABLE

		COND = *1	*1	*2	*2	MARGINAL
		PREF = *1	*2	*1	*2	
	R					
X(3)	1	8.79852	8.89823	8.83945	8.80973	8.82916
X(4)	2	8.79543	8.87855	8.81588	8.81678	8.81971
X(5)	3	8.76835	8.90840	8.84307	8.81674	8.82335
X(6)	4	8.79717	8.86524	8.83282	8.81939	8.82343
X(7)	5	8.79918	8.92574	8.83389	8.79073	8.82828
X(8)	6	8.78005	8.90245	8.81260	8.79533	8.81298
X(9)	7	8.79480	8.88904	8.82897	8.79675	8.82061
X(10)	8	8.78976	8.89995	8.82426	8.79765	8.81961
X(11)	9	8.83166	8.91222	8.86170	8.79077	8.84474
X(12)	10	8.80711	8.89447	8.81452	8.79187	8.82047
X(13)	11	8.77494	8.90357	8.81338	8.78780	8.81005
X(14)	12	8.78964	8.88608	8.82474	8.78289	8.81419
MARGINAL		8.79388	8.89700	8.82877	8.79970	8.82221
COUNT		20	10	17	13	60

STANDARD DEVIATIONS   FOR   1-ST DEPENDENT VARIABLE

		COND = *1	*1	*2	*2
		PREF = *1	*2	*1	*2
	R				
X(3)	1	0.13586	0.14423	0.15743	0.13229
X(4)	2	0.15231	0.12920	0.13535	0.14447
X(5)	3	0.14201	0.14254	0.15475	0.14730
X(6)	4	0.14227	0.16246	0.13637	0.13707
X(7)	5	0.14668	0.13242	0.16442	0.13159
X(8)	6	0.16331	0.12319	0.13134	0.11816
X(9)	7	0.14873	0.10884	0.14223	0.13247
X(10)	8	0.13379	0.12001	0.12929	0.13473
X(11)	9	0.14621	0.13446	0.15650	0.12402
X(12)	10	0.15592	0.12642	0.13875	0.12552
X(13)	11	0.14894	0.13677	0.14040	0.12953
X(14)	12	0.14810	0.10708	0.13829	0.14252

PAGE   4   BMDP2V HAND AND TAYLOR (1987) P.194

```
ANALYSIS OF VARIANCE FOR   1-ST
DEPENDENT VARIABLE - X(3)     X(4)      X(5)      X(6)      X(7)      X(8)
X(9)      X(10)     X(11)
                              X(12)     X(13)     X(14)
```

	SOURCE		SUM OF SQUARES	DEGREES OF FREEDOM	MEAN SQUARE	F
[3]	MEAN		52387.25943	1	52387.25943	242599.27
	COND		0.16354	1	0.16354	0.76
	PREF		0.23022	1	0.23022	1.07
	CP		0.73375	1	0.73375	3.40
1	ERROR		12.09273	56	0.21594	
[4]	R		0.04206	11	0.00382	1.85
	RC		0.03241	11	0.00295	1.42
	RP		0.03083	11	0.00280	1.35
	RCP		0.03202	11	0.00291	1.41
2	ERROR		1.27530	616	0.00207	

	SOURCE	TAIL PROB.	GREENHOUSE GEISSER PROB.	HUYNH FELDT PROB.
[5]	MEAN	0.0000		
	COND	0.3879		
	PREF	0.3063		
	CP	0.0706		
	R	0.0436	0.0853	0.0708
	RC	0.1579	0.2015	0.1884
	RP	0.1907	0.2298	0.2184
	RCP	0.1656	0.2083	0.1956

```
[6]
ERROR                 EPSILON FACTORS FOR DEGREES OF FREEDOM ADJUSTMENT
TERM
                   GREENHOUSE-GEISSER         HUYNH-FELDT
     2                  0.5741                   0.6897
```

----

[3] The anova table. First the results of the analysis based on the
average of the twelve responses. This shows a comparison of the overall
mean with zero (usually not of interest) and the overall COND and PREF
main effects and their interaction. In this case none of these three
latter are significant at the 5% level (see [5] for probabilities).

[4] This part of the table shows the results of the analysis of the
differences in the responses under the twelve conditions. The only
significant difference at the 5% level is that for the overall departure
from equality of the twelve responses (see [5] for probabilities). There
appear to be no differences in shape between the COND levels, the PREF
levels, or the interaction of these two factors.

[5] Probabilities for the anova tests. The first column is the
straightforward test. However, the sphericity test at [2] suggests these
probabilities may be inaccurate. The next two columns show test results
adjusted for the non-sphericity (see Chapter 3). Neither adjustment yields
a test which is significant at the 5% level.

[6] These are the factors used for the sphericity adjustment - see Chapter
3 for explanation.

10.3   Example 2 – BMDP program 2V

Program 2:

BMDP program P2V analysis of the first twelve response variables from Table 10.2

There are four groups arranged in a two by two cross-classification by factors MOTR and PRED. Six exposures to a disturbing stimulus were made and at each exposure skin resistance was measured at onset and at peak value. The onset values are ON1 to ON6 below and the peak values are PK1 to PK6.

```
      /PROBLEM  TITLE='DAUM AND SARTORY DATA, 1988'.
      /INPUT  VARIABLES=14.  UNIT=10.
            FORMAT IS '(4X,F1.0,1X,F1.0,10F6.1/7X,2F6.1//)'.
      /VARIABLE NAMES ARE MOTR,PRED,ON1,PK1,ON2,PK2,ON3,PK3,
[1]                    ON4,PK4,ON5,PK5,ON6,PK6,DF1,DF2,DF3,DF4,DF5,DF6.
[2]            ADD IS 6.
      /TRANSFORM DF1=PK1-ON1.
                 DF2=PK2-ON2.
                 DF3=PK3-ON3.
                 DF4=PK4-ON4.
                 DF5=PK5-ON5.
                 DF6=PK6-ON6.
[3] /DESIGN DEPENDENT IS DF1,DF2,DF3,DF4,DF5,DF6.
            LEVEL IS 6.
            GROUPING ARE MOTR, PRED.
[4]         NAME IS TIME.
      /GROUP CODES(1) ARE 1, 2.
             CODES(2) ARE 1, 2.
      /PRINT LINE=80.
      /END
```

----

[1] DF1 to DF6 are new variables calculated as the difference between peak and onset values for each of the six times as shown in the TRANSFORM paragraph.

[2] Six new variables, DF1 to DF6, are to be added.

[3] The analysis will be on the six difference variables, with subjects cross-classified by MOTR and PRED.

[4] The name 'TIME' has been chosen to describe the within subjects factor with six levels.

Table 10.2. *Data for Examples 2 and 3. There are 48 subject records. The leftmost three digits are a patient identifier. The next two are grouping factors:*

Factor 1, level 1: stimulus switches off when a button is pressed
        level 2: stimulus remains on when the button is pressed
Factor 2, level 1: a tone indicates the stimulus will disappear
        level 2: the tone does not imply this.

Each subject is measured on 36 variables in three sets of 12, the three corresponding to three different stimuli. Within each set the scores are grouped into six pairs, representing six presentations of the stimulus. The first score in each pair is onset level of skin conductance and the second is peak level obtained for this presentation of the stimulus

001 1 1	13.6	15.4	12.3	13.7	12.3	13.6	11.6	12.9	11.7	12.5
	11.4	12.0	10.9	13.1	10.9	12.5	11.1	12.1	10.4	10.4
	10.0	10.0	9.5	9.5	9.9	10.7	9.5	9.5	8.8	8.8
	8.6	8.6	8.3	8.3	8.1	8.1				
002 1 1	16.6	16.6	16.0	16.0	15.8	15.8	15.1	15.1	14.4	14.4
	14.1	14.1	14.0	14.0	13.7	13.7	12.6	12.6	12.6	12.6
	12.5	12.5	12.1	12.7	12.2	12.2	12.1	13.5	11.7	11.7
	11.7	12.4	11.5	12.2	11.1	13.3				
003 1 1	33.7	47.8	34.8	34.8	33.4	35.8	40.3	40.3	34.0	34.0
	36.1	36.1	33.3	38.8	38.2	38.2	38.8	38.8	36.0	36.0
	38.6	38.6	40.5	40.5	35.8	35.8	36.9	36.9	36.0	36.0
	38.2	38.2	37.3	37.3	37.0	37.0				
004 1 1	14.0	14.0	14.3	15.0	13.0	13.7	11.8	12.7	11.8	11.8
	11.5	13.1	11.6	11.6	11.0	11.9	10.6	10.6	10.3	10.7
	9.7	9.7	9.4	10.1	10.5	11.6	9.5	9.5	8.9	8.9
	8.5	8.5	8.3	8.9	8.8	9.0				
005 1 1	6.3	6.5	6.5	6.6	6.2	6.4	6.5	6.5	6.3	6.3
	6.2	6.2	6.0	6.4	5.9	5.9	5.7	5.7	5.3	5.3
	5.0	5.1	4.9	4.9	5.4	5.4	5.2	5.2	4.9	4.9
	4.7	4.7	4.5	4.5	4.4	4.4				
006 1 1	26.2	28.0	27.4	27.4	27.8	29.4	28.7	28.7	29.5	29.5
	31.1	31.1	30.4	30.4	30.5	30.5	30.6	30.6	31.1	31.1
	31.3	31.3	31.5	31.5	31.2	31.2	30.3	30.3	30.9	30.9
	31.0	31.0	31.1	31.1	31.2	31.2				
007 1 1	27.5	27.5	28.7	28.7	28.9	28.9	28.4	28.4	28.3	28.3
	27.2	27.2	26.5	26.5	27.0	27.0	26.4	26.4	25.3	25.3
	24.0	24.0	24.4	24.4	26.6	26.6	27.0	27.0	26.2	26.2
	25.1	25.1	23.5	23.5	22.6	22.6				
008 1 1	4.0	4.3	4.5	4.5	3.9	3.9	4.0	4.0	2.9	2.9
	2.4	2.4	2.2	2.2	2.3	2.3	2.1	2.1	2.1	2.1
	1.9	1.9	2.3	2.3	2.2	2.2	1.8	1.8	1.7	1.7
	2.3	2.3	3.3	3.3	2.8	2.8				
009 1 1	10.0	12.6	10.8	11.3	10.1	10.1	9.3	9.3	8.6	8.6
	8.2	8.4	8.8	10.2	10.2	10.2	9.1	9.7	8.9	9.0
	8.7	8.7	8.5	8.5	9.0	10.0	9.1	9.6	8.7	8.7
	8.1	8.1	8.3	8.3	9.4	9.6				
010 1 1	6.1	7.0	6.5	6.8	5.9	5.9	5.4	5.4	4.7	4.7
	4.4	4.4	4.0	4.0	3.7	3.7	3.6	3.6	3.3	3.3
	3.2	3.2	2.9	2.9	3.6	3.6	3.2	3.2	2.9	3.4
	2.7	2.7	2.6	2.6	2.5	2.5				
011 1 1	9.6	10.3	9.2	9.2	9.1	9.1	8.8	8.8	8.4	8.4
	8.1	8.1	7.9	7.9	8.1	8.1	7.8	7.8	7.6	7.6
	7.4	7.4	7.4	7.4	7.4	7.4	7.2	7.2	7.0	7.0
	6.9	6.9	6.8	6.8	6.6	6.6				
012 1 1	2.7	3.2	3.6	3.9	5.3	6.5	6.0	6.9	5.4	5.7
	4.9	4.9	4.3	5.1	4.4	4.9	4.2	4.2	3.6	3.6
	3.1	3.1	2.8	2.8	3.1	3.5	3.1	3.1	2.7	2.7
	2.5	2.5	2.3	2.3	2.2	2.2				

(Contd.)

Table 10.2. *(Contd.)*

013	1	2	11.7	20.6	16.5	22.5	14.5	14.5	15.1	15.1	16.8	20.4
			14.4	14.4	14.9	14.9	14.5	14.5	14.7	14.7	13.2	14.8
			11.4	11.4	10.5	10.5	15.6	15.6	12.2	12.2	10.2	10.2
			8.6	8.6	11.5	11.5	9.7	9.7				
014	1	2	32.1	32.1	32.4	32.4	31.7	31.7	33.1	33.1	33.4	33.4
			33.6	33.6	34.5	34.5	34.7	34.7	33.6	33.6	33.7	33.7
			34.0	34.0	35.7	35.7	35.3	35.3	34.6	34.6	34.5	34.5
			36.4	36.4	35.1	35.1	36.6	36.6				
015	1	2	20.0	21.0	20.4	20.4	20.4	20.4	20.6	20.6	21.0	21.6
			20.8	20.8	21.5	21.5	20.7	20.7	21.2	22.3	21.9	21.9
			21.6	21.6	23.0	23.0	21.6	21.6	21.8	21.8	21.9	23.5
			22.4	22.4	22.4	24.0	22.5	22.5				
016	1	2	4.1	4.2	4.1	4.1	4.0	4.0	3.8	3.8	3.6	3.6
			3.5	3.5	3.3	3.3	3.2	3.2	3.1	3.1	3.0	3.0
			2.9	2.9	2.8	2.8	2.6	2.6	2.5	2.5	2.5	2.5
			2.4	2.4	2.4	2.4	2.4	2.4				
017	1	2	6.0	7.8	5.5	6.6	4.9	5.1	4.6	4.6	4.1	4.1
			3.4	3.4	3.0	3.3	5.8	6.5	6.4	6.7	7.2	7.6
			5.4	5.4	4.9	4.9	6.1	6.8	5.1	5.1	4.0	4.0
			3.4	3.4	3.2	3.2	2.9	2.9				
018	1	2	9.5	10.1	9.2	9.9	9.0	9.0	8.5	8.5	8.3	8.9
			8.2	8.6	8.1	8.7	8.0	8.5	7.7	7.7	7.5	8.3
			7.4	7.4	6.9	6.9	6.8	7.3	6.7	6.7	6.7	6.7
			6.3	6.3	5.7	5.7	5.2	7.0				
019	1	2	2.1	2.1	1.9	1.9	1.8	1.8	1.7	1.7	1.6	1.6
			1.6	1.6	1.6	1.6	1.6	1.6	1.8	1.8	1.5	1.6
			1.9	1.9	2.7	2.7	2.1	2.1	2.1	2.1	1.8	1.8
			1.8	1.8	1.7	1.7	1.7	1.7				
020	1	2	21.9	22.9	21.4	24.6	21.9	23.7	22.1	24.0	21.8	23.8
			21.0	24.3	21.1	23.7	23.0	24.9	22.9	24.4	23.9	23.9
			24.9	24.9	24.6	24.6	27.0	29.3	25.1	26.4	26.6	26.6
			25.8	25.8	24.4	25.6	24.5	24.5				
021	1	2	5.0	8.0	6.4	8.3	5.5	5.5	5.5	5.7	6.0	6.0
			5.3	5.3	5.2	5.5	5.2	5.2	5.2	5.2	5.2	5.8
			5.2	5.2	5.1	5.1	6.0	6.2	5.2	5.2	4.8	4.8
			4.5	4.5	4.3	4.3	4.1	4.1				
022	1	2	7.6	9.1	8.7	12.0	9.3	11.2	9.4	11.9	9.8	11.6
			10.2	12.2	9.6	14.7	10.5	12.3	10.1	11.5	9.8	12.3
			9.6	11.3	9.6	10.2	9.6	12.6	10.0	10.0	9.6	9.6
			9.5	9.5	9.0	9.0	9.1	9.1				
023	1	2	24.3	29.6	25.0	25.0	22.9	22.9	20.7	20.7	19.3	19.8
			18.5	18.5	17.6	17.6	17.7	17.7	16.5	16.5	15.6	15.6
			15.5	15.5	15.6	15.6	20.9	21.9	20.9	20.9	18.7	18.7
			17.5	17.5	16.8	16.8	16.2	16.2				
024	1	2	6.9	7.6	7.9	8.4	7.3	7.3	7.1	7.4	7.4	7.4
			7.1	7.1	6.9	6.9	7.0	7.2	7.6	7.7	7.4	7.4
			7.1	7.2	7.1	7.1	8.1	8.3	7.9	7.9	7.8	7.8
			8.5	8.5	7.8	7.8	7.7	7.7				
025	2	1	12.4	16.3	12.6	14.8	12.3	14.5	11.6	13.3	12.0	13.1
			11.4	12.5	10.9	14.2	12.4	13.7	11.6	12.5	11.5	11.5
			11.3	11.3	10.3	10.3	13.9	16.8	12.8	13.9	12.3	13.2
			11.5	13.3	11.3	12.7	11.5	12.7				
026	2	1	21.7	21.7	22.1	22.1	22.3	24.3	22.7	22.7	22.7	22.7
			22.8	22.8	23.1	23.1	24.5	24.5	23.1	23.1	22.1	22.1
			23.9	23.9	23.0	23.0	24.6	24.6	25.4	25.4	23.0	24.3
			21.8	21.8	22.4	22.4	22.1	22.6				
027	2	1	25.0	25.0	24.0	26.0	24.3	25.5	23.4	25.0	24.8	24.8
			24.8	24.8	25.4	25.4	25.8	25.8	27.5	27.5	28.4	28.4
			28.4	28.4	26.5	26.5	26.5	28.3	26.5	29.0	27.0	27.0
			24.4	27.5	26.5	26.5	24.4	24.4				

(Contd.)

Table 10.2. *(Contd.)*

028 2 1	2.7	3.1	2.8	2.8	2.6	2.6	2.5	2.5	2.4	2.4
	2.3	2.3	2.4	2.4	2.1	2.1	2.1	2.1	2.0	2.0
	2.0	2.0	1.9	1.9	4.1	4.1	3.6	3.6	3.1	3.1
	2.8	2.8	2.5	2.5	2.4	2.4				
029 2 1	18.4	18.4	17.4	17.4	16.5	16.5	9.7	10.1	10.0	10.0
	9.8	9.8	9.8	9.8	9.6	9.6	9.5	9.5	9.4	9.4
	9.3	9.3	9.3	9.3	10.0	10.0	9.5	9.5	9.1	9.1
	8.9	8.9	8.8	8.8	8.8	8.8				
030 2 1	11.8	14.9	12.2	14.4	11.1	13.4	10.4	11.9	9.7	10.1
	8.9	8.9	8.3	10.7	9.0	9.0	9.5	11.7	9.3	9.3
	7.9	7.9	9.3	9.3	11.4	14.6	10.7	10.7	9.2	9.2
	9.0	9.0	9.1	11.5	9.8	9.8				
031 2 1	2.5	3.1	2.6	3.1	2.9	3.7	3.2	3.7	3.6	4.8
	3.6	4.4	3.8	4.8	3.5	3.5	3.6	3.6	3.4	3.4
	2.7	2.7	2.5	2.5	3.4	3.7	3.5	3.5	3.1	3.1
	2.8	2.8	2.7	2.7	2.5	2.5				
032 2 1	11.8	13.2	14.0	15.0	12.7	13.6	11.4	12.4	10.5	12.1
	10.0	10.4	9.1	10.2	8.7	9.1	8.0	8.1	7.3	7.7
	10.4	10.7	9.6	9.8	8.1	10.1	6.7	6.7	6.1	6.2
	5.6	5.6	5.4	5.4	5.0	5.0				
033 2 1	5.7	7.1	6.6	6.6	5.5	5.7	5.3	5.3	5.1	5.2
	5.4	5.4	5.2	5.6	5.2	7.3	6.8	6.9	6.3	6.3
	6.1	6.1	5.4	5.4	5.2	5.2	5.3	5.3	4.9	4.9
	5.1	5.1	5.0	6.5	7.0	7.0				
034 2 1	11.2	11.2	11.1	11.6	11.4	11.8	11.5	11.5	11.7	11.7
	11.5	11.5	11.8	12.9	12.6	13.6	12.9	12.9	12.9	12.9
	12.7	12.7	13.4	13.4	14.3	15.9	13.6	13.6	13.0	13.0
	13.1	13.1	13.3	13.3	14.2	14.2				
035 2 1	28.2	33.3	28.0	30.6	27.0	27.0	26.5	26.5	25.1	29.1
	24.5	24.5	23.3	23.3	23.3	23.3	23.9	23.9	24.5	24.5
	22.9	22.9	21.6	21.6	28.0	28.0	23.8	23.8	21.6	21.6
	19.5	19.5	19.5	20.0	19.6	19.6				
036 2 1	9.1	10.4	9.5	10.2	9.7	10.8	9.6	10.1	9.4	10.4
	9.5	9.8	9.0	9.5	9.1	9.6	8.9	9.6	9.0	9.4
	8.4	8.4	8.4	8.9	8.7	9.5	9.8	10.4	8.7	8.7
	8.1	8.1	8.2	8.2	8.4	8.4				
037 2 2	22.4	22.4	23.7	23.7	23.2	23.2	23.9	23.9	23.0	23.0
	22.6	22.6	22.9	22.9	22.3	22.3	21.6	21.6	20.5	20.5
	20.3	20.3	19.6	19.6	20.6	20.6	20.1	20.1	19.8	19.8
	19.4	19.4	18.9	18.9	17.8	17.8				
038 2 2	15.0	15.8	15.0	6.6	14.9	16.8	17.0	18.0	15.5	17.3
	16.3	18.8	18.2	19.5	15.3	15.3	16.2	16.2	15.3	16.2
	15.7	16.5	14.7	15.4	17.1	19.0	15.0	16.1	13.5	13.5
	12.7	12.7	12.0	12.0	12.5	12.5				
039 2 2	1.9	1.9	2.0	2.0	2.0	2.0	2.0	2.0	1.9	1.9
	1.9	1.9	2.1	2.1	2.1	2.1	2.0	2.0	2.0	2.0
	2.0	2.0	2.0	2.0	1.9	1.9	1.9	1.9	1.9	1.9
	2.1	2.1	2.1	2.1	2.1	2.1				
040 2 2	2.6	2.7	2.7	2.7	2.7	2.8	2.4	2.4	2.3	2.3
	2.2	2.2	2.1	2.1	2.0	2.0	2.0	2.0	1.9	1.9
	1.9	1.9	1.9	1.9	1.9	1.9	1.8	1.8	1.8	1.8
	1.9	1.9	1.8	1.8	1.8	1.8				
041 2 2	14.5	14.5	13.8	14.6	13.3	13.3	12.8	12.8	12.5	12.5
	12.3	12.3	12.1	12.1	12.0	12.0	11.4	11.4	10.9	10.9
	10.4	10.4	9.7	9.7	13.4	14.1	12.5	12.5	11.1	11.3
	10.0	10.0	9.2	9.2	8.5	14.2				
042 2 2	1.7	1.9	1.7	1.7	1.6	1.6	1.5	1.5	1.3	1.3
	1.4	1.4	1.4	1.4	1.3	1.4	1.3	1.3	1.3	1.3
	1.3	1.3	1.3	1.3	1.3	1.3	1.4	1.4	1.4	1.4
	1.4	1.4	1.4	1.4						

(Contd.)

Table 10.2. *(Contd.)*

043 2 2	3.5	3.6	3.3	3.3	3.0	3.0	2.7	2.7	2.5	2.5
	2.4	2.4	2.2	2.2	2.2	2.2	2.1	2.1	2.0	2.0
	2.0	2.0	1.9	1.9	2.2	2.2	2.0	2.0	1.9	1.9
	2.8	2.8	2.7	2.7	2.6	2.6				
044 2 2	1.9	1.9	1.9	1.9	1.9	1.9	1.9	1.9	1.9	1.9
	2.0	2.0	2.0	2.0	2.0	2.0	2.0	2.0	2.0	2.0
	2.1	2.1	2.1	2.1	2.0	2.0	2.0	2.0	2.0	2.0
	2.0	2.0	2.0	2.0	2.0	2.0				
045 2 2	8.7	10.0	7.0	7.2	6.4	6.4	6.1	6.1	6.2	6.7
	5.7	5.7	5.5	5.7	5.4	5.4	5.2	5.2	5.1	5.1
	5.1	5.1	5.0	5.3	5.3	5.3	5.2	5.2	5.1	5.1
	5.0	5.0	4.9	4.9	4.9	4.9				
046 2 2	18.3	20.0	18.1	19.0	16.4	17.1	15.6	16.7	15.3	15.3
	14.6	14.6	13.4	17.2	13.6	14.1	12.7	12.7	12.2	12.2
	11.4	11.7	11.1	12.0	13.1	15.5	12.5	13.5	12.0	12.6
	11.7	12.9	11.1	11.1	10.6	10.6				
047 2 2	5.7	6.1	4.7	4.8	4.6	4.9	5.2	5.4	5.5	5.9
	5.7	5.9	5.8	6.7	5.6	6.1	6.0	6.9	6.6	7.5
	6.3	6.6	6.3	6.8	5.6	6.0	4.8	5.4	5.4	5.6
	5.3	5.9	4.9	5.0	4.7	5.1				
048 2 2	15.8	19.5	10.1	11.5	9.6	10.4	9.4	10.5	9.2	9.2
	8.7	8.9	8.4	10.3	8.9	8.9	8.2	8.2	7.8	7.8
	7.3	7.3	7.2	7.2	8.8	8.8	8.1	8.1	7.5	7.5
	7.3	7.3	7.1	7.1	6.9	6.9				

This data set was kindly supplied by Drs Irene Daun and Gudrun Sartory of the London University Institute of Psychiatry.

Output from program 2:

This has a structure similar to that of the output from program 1. In this case, however, some of the adjusted effects are significant.

PROBLEM TITLE IS
DAUM AND SARTORY DATA, 1988

NUMBER OF VARIABLES TO READ IN.   14
NUMBER OF VARIABLES ADDED BY TRANSFORMATIONS. .   6
TOTAL NUMBER OF VARIABLES   20
NUMBER OF CASES TO READ IN. . . . . TO END
CASE LABELING VARIABLES . . . .
MISSING VALUES CHECKED BEFORE OR AFTER TRANS. . NEITHER
BLANKS ARE. . . . . MISSING
NPUT UNIT NUMBER   10
REWIND INPUT UNIT PRIOR TO READING. . DATA. . .   YES
NUMBER OF WORDS OF DYNAMIC STORAGE.   351230
NUMBER OF CASES DESCRIBED BY INPUT FORMAT . . .   1

VARIABLES TO BE USED
      1 MOTR   2 PRED   3 ON1   4 PK1   5 ON2
      6 PK2   7 ON3   8 PK3   9 ON4   10 PK4
     11 ON5   12 PK5   13 ON6   14 PK6   15 DF1
     16 DF2   17 DF3   18 DF4   19 DF5   20 DF6

INPUT FORMAT IS
(4X,F1.0,1X,F1.0,10F6.1/7X,2F6.1//)

MAXIMUM LENGTH DATA RECORD IS   67 CHARACTERS.

I N P U T   V A R I A B L E S . . . . .
  VARIABLE RECORD COLUMN   INPUT   VARIABLE RECORD COLUMN   INPUT
  NO. NAME   NO. BEG END FORMAT   NO. NAME   NO. BEG END FORMAT
  --- ------ --- --- --- ------   --- ------ --- --- --- ------
   1 MOTR   1   5   5   F1.0   8 PK3   1   38   43   F6.1
   2 PRED   1   7   7   F1.0   9 ON4   1   44   49   F6.1
   3 ON1   1   8   13   F6.1   10 PK4   1   50   55   F6.1
   4 PK1   1   14   19   F6.1   11 ON5   1   56   61   F6.1
   5 ON2   1   20   25   F6.1   12 PK5   1   62   67   F6.1
   6 PK2   1   26   31   F6.1   13 ON6   2   8   13   F6.1
   7 ON3   1   32   37   F6.1   14 PK6   2   14   19   F6.1

DESIGN SPECIFICATIONS

          GROUP =   1   2
          DEPEND = 15   16   17   18   19   20
          LEVEL =   6

BASED ON INPUT FORMAT SUPPLIED   4 RECORDS READ PER CASE.

*** WARNING ---   2 EXTRA RECORDS AT END OF EACH CASE ***

VARIABLE		STATED VALUES FOR				GROUP	CATEGORY	INTERVALS	
NO.	NAME	MINIMUM	MAXIMUM	MISSING	CODE	INDEX	NAME	.GT.	.LE.
1	MOTR				1.000	1	*1		
					2.000	2	*2		
2	PRED				1.000	1	*1		
					2.000	2	*2		

NOTE: CATEGORY NAMES BEGINNING WITH * WERE CREATED BY THE PROGRAM.

NUMBER OF CASES READ.   48
PAGE   2   BMDP2V DAUM AND SARTORY DATA, 1988

GROUP STRUCTURE

MOTR	PRED	COUNT
*1	*1	12
*1	*2	12
*2	*1	12
*2	*2	12

SUMS OF SQUARES AND CORRELATION MATRIX OF THE
ORTHOGONAL COMPONENTS POOLED FOR ERROR   2   IN ANOVA TABLE BELOW

```
        149.44442      1.000
         69.89177      0.775  1.000
         57.82116      0.438  0.833  1.000
         61.08017     -0.169  0.170  0.414  1.000
         24.17196      0.038  0.503  0.784  0.687  1.000
[1]
```
SPHERICITY TEST APPLIED TO ORTHOGONAL COMPONENTS - TAIL PROBABILITY   0.0000

CELL MEANS FOR   1-ST DEPENDENT VARIABLE

		MOTR = *1	*1	*2	*2	MARGINAL
		PRED = *1	*2	*1	*2	
	TIME					
DF1	1	1.90833	1.99167	1.43333	0.69167	1.50625
DF2	2	0.27500	1.39167	0.97500	-0.41667	0.55625
DF3	3	0.61667	0.32500	0.90833	0.31667	0.54167
DF4	4	0.25833	0.40833	0.60000	0.28333	0.38750
DF5	5	0.09167	0.75833	0.78333	0.22500	0.46458
DF6	6	0.20000	0.47500	0.21667	0.24167	0.28333
	MARGINAL	0.55833	0.89167	0.81944	0.22361	0.62326
	COUNT	12	12	12	12	48

---

[1] The sphericity test shows that the unadjusted univariate approach to
repeated measures is not valid (see Chapter 3).

STANDARD DEVIATIONS FOR 1-ST DEPENDENT VARIABLE

		MOTR	=	*1	*1	*2	*2
		PRED	=	*1	*2	*1	*2
		TIME					

DF1	1	3.93156	2.64659	1.71376	1.10327
DF2	2	0.42453	1.89375	1.00102	2.55728
DF3	3	0.82112	0.71494	0.87849	0.57498
DF4	4	0.47760	0.85223	0.67689	0.47641
DF5	5	0.23916	1.13895	1.16840	0.52592
DF6	6	0.47482	1.05926	0.37376	0.71536

PAGE 3 BMDP2V DAUM AND SARTORY DATA, 1988

ANALYSIS OF VARIANCE FOR 1-ST
DEPENDENT VARIABLE - DF1 DF2 DF3 DF4 DF5 DF6

	SOURCE	SUM OF SQUARES	DEGREES OF FREEDOM	MEAN SQUARE	F
	MEAN	111.87581	1	111.87581	32.19
	MOTR	2.98087	1	2.98087	0.86
	PRED	1.24031	1	1.24031	0.36
	MP	15.54031	1	15.54031	4.47
1	ERROR	152.90087	44	3.47502	
	TIME	47.38223	5	9.47645	5.75
	TM	10.75391	5	2.15078	1.31
	TP	3.01614	5	0.60323	0.37
	TMP	10.98945	5	2.19789	1.33
2	ERROR	362.40948	220	1.64732	

	SOURCE	TAIL PROB.	GREENHOUSE GEISSER PROB.	HUYNH FELDT PROB.
	MEAN	0.0000		
	MOTR	0.3594		
	PRED	0.5533		
	MP	0.0402		
2]	TIME	0.0001	0.0038	0.0026
	TM	0.2625	0.2765	0.2768
	TP	0.8714	0.7054	0.7295
	TMP	0.2508	0.2688	0.2686

RROR ERM	EPSILON FACTORS FOR DEGREES OF FREEDOM ADJUSTMENT	
	GREENHOUSE-GEISSER	HUYNH-FELDT
2	0.4216	0.4735

2] The TIME effect is significant when either adjustment is made (see
napter 3).

10.4   Example 3 – BMDP program 2V

Program 3:

BMDP program P2V analysis of the data from Table 10.2. The data analysed
in example 2 are a subset of these data.

```
/PROBLEM  TITLE='DAUM AND SARTORY DATA, 1988'.
/INPUT    VARIABLES=38.   UNIT=10.
          FORMAT IS '(4X,F1.0,1X,F1.0,10F6.1/7X,10F6.1/7X,10F6.1/7X,6F6.1)'.
/VARIABLE NAMES ARE MOTR,PRED,
                    OL1,PL1,OL2,PL2,OL3,PL3,OL4,PL4,OL5,PL5,OL6,PL6,
                    OM1,PM1,OM2,PM2,OM3,PM3,OM4,PM4,OM5,PM5,OM6,PM6,
                    ON1,PN1,ON2,PN2,ON3,PN3,ON4,PN4,ON5,PN5,ON6,PN6,
[1]                 DL1,DL2,DL3,DL4,DL5,DL6,
                    DM1,DM2,DM3,DM4,DM5,DM6,
                    DN1,DN2,DN3,DN4,DN5,DN6.
          ADD IS 18.
/TRANSFORM DL1=PL1-OL1.
           DL2=PL2-OL2.
           DL3=PL3-OL3.
           DL4=PL4-OL4.
           DL5=PL5-OL5.
           DL6=PL6-OL6.
           DM1=PM1-OM1.
           DM2=PM2-OM2.
           DM3=PM3-OM3.
           DM4=PM4-OM4.
           DM5=PM5-OM5.
           DM6=PM6-OM6.
           DN1=PN1-ON1.
           DN2=PN2-ON2.
           DN3=PN3-ON3.
           DN4=PN4-ON4.
           DN5=PN5-ON5.
           DN6=PN6-ON6.
/DESIGN DEPENDENT IS DL1,DL2,DL3,DL4,DL5,DL6,
                     DM1,DM2,DM3,DM4,DM5,DM6,
                     DN1,DN2,DN3,DN4,DN5,DN6.
[2]        LEVEL IS 3, 6.
           GROUPING ARE MOTR, PRED.
           NAMES ARE SLIDE, TIME.
/GROUP CODES(1) ARE 1, 2.
       CODES(2) ARE 1, 2.
/PRINT LINE=80.
/END
```

---

[1] The analysis is again based on differences. Now there are 18, defined
as in the TRANSFORM paragraph.

[2] The measurements are organised into three consecutive blocks of six.
Thus within subjects there is a three by six factorial arrangement of
difference measurements.

Output from program 3:

BMDP program P2V analysis of a repeated measures design in which responses
are measured at eighteen times, arranged in three blocks of six. Thus we
have a three (SLIDE) by six (TIME) design within each subject. In fact two
measures are taken at each of the eighteen times - onset and peak values -
but the analysis is based on the difference between these two scores.

```
PROBLEM TITLE IS
DAUM AND SARTORY DATA, 1988

NUMBER OF VARIABLES TO READ IN. . . . . . . . .      38
NUMBER OF VARIABLES ADDED BY TRANSFORMATIONS. .      18
TOTAL NUMBER OF VARIABLES . . . . . . . . . . .      56
NUMBER OF CASES TO READ IN. . . . . . . . . . . TO END
CASE LABELING VARIABLES . . . . . . . . . . . .
MISSING VALUES CHECKED BEFORE OR AFTER TRANS. . NEITHER
BLANKS ARE. . . . . . . . . . . . . . . . . . . MISSING
INPUT UNIT NUMBER . . . . . . . . . . . . . . .      10
REWIND INPUT UNIT PRIOR TO READING. . DATA. . .     YES
NUMBER OF WORDS OF DYNAMIC STORAGE. . . . . . .  351230
NUMBER OF CASES DESCRIBED BY INPUT FORMAT . . .       1

VARIABLES TO BE USED
      1 MOTR       2 PRED       3 OL1       4 PL1       5 OL2
      6 PL2        7 OL3        8 PL3       9 OL4      10 PL4
     11 OL5       12 PL5       13 OL6      14 PL6      15 OM1
     16 PM1       17 OM2       18 PM2      19 OM3      20 PM3
     21 OM4       22 PM4       23 OM5      24 PM5      25 OM6
     26 PM6       27 ON1       28 PN1      29 ON2      30 PN2
     31 ON3       32 PN3       33 ON4      34 PN4      35 ON5
     36 PN5       37 ON6       38 PN6      39 DL1      40 DL2
     41 DL3       42 DL4       43 DL5      44 DL6      45 DM1
     46 DM2       47 DM3       48 DM4      49 DM5      50 DM6
     51 DN1       52 DN2       53 DN3      54 DN4      55 DN5
     56 DN6

INPUT FORMAT IS
(4X,F1.0,1X,F1.0,10F6.1/7X,10F6.1/7X,10F6.1/7X,6F6.1)

MAXIMUM LENGTH DATA RECORD IS    67 CHARACTERS.
```

```
I N P U T    V A R I A B L E S . . . . .
```

VARIABLE NO.	NAME	RECORD NO.	COLUMN BEG	END	INPUT FORMAT	VARIABLE NO.	NAME	RECORD NO.	COLUMN BEG	END	INPUT FORMAT
1	MOTR	1	5	5	F1.0	20	PM3	2	50	55	F6.1
2	PRED	1	7	7	F1.0	21	OM4	2	56	61	F6.1
3	OL1	1	8	13	F6.1	22	PM4	2	62	67	F6.1
4	PL1	1	14	19	F6.1	23	OM5	3	8	13	F6.1
5	OL2	1	20	25	F6.1	24	PM5	3	14	19	F6.1
6	PL2	1	26	31	F6.1	25	OM6	3	20	25	F6.1
7	OL3	1	32	37	F6.1	26	PM6	3	26	31	F6.1
8	PL3	1	38	43	F6.1	27	ON1	3	32	37	F6.1
9	OL4	1	44	49	F6.1	28	PN1	3	38	43	F6.1
10	PL4	1	50	55	F6.1	29	ON2	3	44	49	F6.1

11	OL5	1	56	61	F6.1	30	PN2	3	50	55	F6.1
12	PL5	1	62	67	F6.1	31	ON3	3	56	61	F6.1
13	OL6	2	8	13	F6.1	32	PN3	3	62	67	F6.1
14	PL6	2	14	19	F6.1	33	ON4	4	8	13	F6.1
15	OM1	2	20	25	F6.1	34	PN4	4	14	19	F6.1
16	PM1	2	26	31	F6.1	35	ON5	4	20	25	F6.1
17	OM2	2	32	37	F6.1	36	PN5	4	26	31	F6.1
18	PM2	2	38	43	F6.1	37	ON6	4	32	37	F6.1
19	OM3	2	44	49	F6.1	38	PN6	4	38	43	F6.1

DESIGN SPECIFICATIONS

```
        GROUP =   1   2
        DEPEND = 39  40  41  42  43  44  45  46  47  48  49  50  51  52  53
                 54  55  56
        LEVEL =   3   6
```

BASED ON INPUT FORMAT SUPPLIED   4 RECORDS READ PER CASE.

VARIABLE		STATED VALUES FOR				GROUP	CATEGORY	INTERVALS	
NO.	NAME	MINIMUM	MAXIMUM	MISSING	CODE	INDEX	NAME	.GT.	.LE.
1	MOTR				1.000	1	*1		
					2.000	2	*2		
2	PRED				1.000	1	*1		
					2.000	2	*2		

NOTE: CATEGORY NAMES BEGINNING WITH * WERE CREATED BY THE PROGRAM.

```
NUMBER OF CASES READ. . . . . . . . . . . . . .      48
     CASES WITH DATA MISSING OR BEYOND LIMITS . .       1
          REMAINING NUMBER OF CASES . . . . . . . .     47
PAGE   3  BMDP2V DAUM AND SARTORY DATA, 1988
```

GROUP STRUCTURE

MOTR	PRED	COUNT
*1	*1	12
*1	*2	12
*2	*1	12
*2	*2	11

SUMS OF SQUARES AND CORRELATION MATRIX OF THE
ORTHOGONAL COMPONENTS POOLED FOR ERROR   2 IN ANOVA TABLE BELOW

```
        75.57582        1.000
        22.85407        0.344   1.000
```

SPHERICITY TEST APPLIED TO ORTHOGONAL COMPONENTS - TAIL PROBABILITY   0.0001

SUMS OF SQUARES AND CORRELATION MATRIX OF THE
ORTHOGONAL COMPONENTS POOLED FOR ERROR   3 IN ANOVA TABLE BELOW

```
        91.13511        1.000
        46.47218        0.716   1.000
        38.04292        0.634   0.723   1.000
        25.05079        0.214   0.431   0.528   1.000
        18.27226        0.104   0.320   0.515   0.509   1.000
```

SPHERICITY TEST APPLIED TO ORTHOGONAL COMPONENTS - TAIL PROBABILITY   0.0000

SUMS OF SQUARES AND CORRELATION MATRIX OF THE
ORTHOGONAL COMPONENTS POOLED FOR ERROR   4 IN ANOVA TABLE BELOW

```
        95.46636        1.000
        48.20890        0.624   1.000
        34.85418        0.423   0.671   1.000
        34.65160       -0.290   0.181   0.251   1.000
        14.25215       -0.052   0.356   0.682   0.570   1.000
        23.77345        0.574   0.632   0.395  -0.269   0.139   1.000
        13.87201        0.439   0.308   0.517  -0.184   0.114   0.420   1.000
        10.60925        0.033   0.471   0.369   0.355   0.235   0.249   0.415
        12.54757       -0.382  -0.103   0.198   0.688   0.472  -0.375  -0.217
         9.67545        0.019   0.355   0.608   0.288   0.391   0.086   0.407
```

PAGE   4   BMDP2V DAUM AND SARTORY DATA, 1988

```
        10.60925        1.000
        12.54757        0.084   1.000
         9.67545        0.364   0.386   1.000
```
[1]
SPHERICITY TEST APPLIED TO ORTHOGONAL COMPONENTS - TAIL PROBABILITY   0.0000

         CELL MEANS FOR   1-ST DEPENDENT VARIABLE

		MOTR = *1	*1	*2	*2
		PRED = *1	*2	*1	*2
SLID	TIME				
E					
DL1	1	1	1.90833	1.99167	1.43333	0.73636
DL2	1	2	0.27500	1.39167	0.97500	-0.45454

----

[1] The sphericity test shows that an unadjusted univariate analysis of
variance approach cannot be justified. However, this test is based on all
eighteen variables at once. As discussed in Chapter 3, this is not of
primary interest. In fact, new variables, derived by combining and
transforming the old ones, fall into natural groups and it is sphericity
within each of the groups which is of interest.

DL3	1	3	0.61667	0.32500	0.90833	0.34545
DL4	1	4	0.25833	0.40833	0.60000	0.30909
DL5	1	5	0.09167	0.75833	0.78333	0.24545
DL6	1	6	0.20000	0.47500	0.21667	0.26364
DM1	2	1	0.85833	0.74167	0.81667	0.73636
DM2	2	2	0.25000	0.42500	0.44167	0.09091
DM3	2	3	0.13333	0.36667	0.33333	0.08182
DM4	2	4	0.04167	0.50000	0.06667	0.16364
DM5	2	5	0.00833	0.15000	0.02500	0.12727
DM6	2	6	0.10833	0.05000	0.05833	0.20000
DN1	3	1	0.27500	0.65833	1.05000	0.51818
DN2	3	2	0.15833	0.10833	0.35000	0.24545
DN3	3	3	0.04167	0.13333	0.19167	0.09091
DN4	3	4	0.05833	0.00000	0.40833	0.16364
DN5	3	5	0.10833	0.23333	0.48333	0.00909
DN6	3	6	0.21667	0.15000	0.14167	0.55455
MARGINAL			0.31157	0.49259	0.51574	0.24596
COUNT			12	12	12	11

MARGINAL

SLID TIME
E

DL1	1	1	1.53404
DL2	1	2	0.56809
DL3	1	3	0.55319
DL4	1	4	0.39574
DL5	1	5	0.47447
DL6	1	6	0.28936
DM1	2	1	0.78936
DM2	2	2	0.30638
DM3	2	3	0.23191
DM4	2	4	0.19362
DM5	2	5	0.07660
DM6	2	6	0.10213
DN1	3	1	0.62766
DN2	3	2	0.21489
DN3	3	3	0.11489
DN4	3	4	0.15745
DN5	3	5	0.21277
DN6	3	6	0.25957
MARGINAL			0.39456
COUNT			47

PAGE 5 BMDP2V DAUM AND SARTORY DATA, 1988

STANDARD DEVIATIONS FOR 1-ST DEPENDENT VARIABLE

			MOTR = *1	*1	*2	*2
			PRED = *1	*2	*1	*2
	SLID	TIME E				
DL1	1	1	3.93156	2.64659	1.71376	1.14567
DL2	1	2	0.42453	1.89375	1.00102	2.67857
DL3	1	3	0.82112	0.71494	0.87849	0.59391

```
DL4        1    4    0.47760    0.85223    0.67689    0.49082
DL5        1    5    0.23916    1.13895    1.16840    0.54656
DL6        1    6    0.47482    1.05926    0.37376    0.74602
DM1        2    1    1.62394    1.55824    1.06501    1.20936
DM2        2    2    0.51079    0.70469    0.68551    0.20226
DM3        2    3    0.32287    0.59594    0.66241    0.27136
DM4        2    4    0.11645    0.79658    0.15570    0.36407
DM5        2    5    0.02887    0.48897    0.08660    0.25334
DM6        2    6    0.25391    0.17321    0.15050    0.33466
DN1        3    1    0.43719    0.99586    1.20416    0.91084
DN2        3    2    0.41661    0.37528    0.75978    0.43672
DN3        3    3    0.14434    0.46188    0.43371    0.18684
DN4        3    4    0.20207    0.00000    0.99312    0.38800
DN5        3    5    0.25391    0.55158    0.82112    0.03015
DN6        3    6    0.62933    0.51962    0.36296    1.71077
PAGE   6  BMDP2V DAUM AND SARTORY DATA, 1988
```

ANALYSIS OF VARIANCE FOR   1-ST
DEPENDENT VARIABLE - DL1   DL2   DL3   DL4   DL5   DL6
DM1   DM2   DM3
                                DM4   DM5   DM6   DN1   DN2   DN3
DN4   DN5   DN6

[2]	SOURCE	SUM OF SQUARES	DEGREES OF FREEDOM	MEAN SQUARE	F
	MEAN	129.46233	1	129.46233	40.53
	MOTR	0.09522	1	0.09522	0.03
	PRED	0.41600	1	0.41600	0.13
	MP	10.73003	1	10.73003	3.36
1	ERROR	137.36405	43	3.19451	
	SLIDE	23.59471	2	11.79735	10.31
	SM	4.78010	2	2.39005	2.09
	SP	0.94719	2	0.47360	0.41
	SMP	5.63325	2	2.81662	2.46
2	ERROR	98.42989	86	1.14453	
	TIME	59.50565	5	11.90113	11.69
	TM	2.54099	5	0.50820	0.50
	TP	2.21311	5	0.44262	0.43
	TMP	6.87572	5	1.37514	1.35
3	ERROR	218.97326	215	1.01848	
	ST	11.33440	10	1.13344	1.64
	STM	8.97615	10	0.89761	1.30
	STP	2.78698	10	0.27870	0.40
	STMP	9.17195	10	0.91720	1.32
4	ERROR	297.91093	430	0.69282	

2] The anova table is divided into four sections, one for the overall
ean of the eighteen response variables, one for variables representing
ifferences between the blocks of times (SLIDE), one for variables
epresenting differences over the six times within each block (TIME), and
ne for the interaction between the blocks and times (the SLIDE by TIME
nteraction).

	SOURCE	TAIL PROB.	GREENHOUSE GEISSER PROB.	HUYNH FELDT PROB.
	MEAN	0.0000		
	MOTR	0.8637		
	PRED	0.7200		
	MP	0.0738		
[3]	SLIDE	0.0001	0.0006	0.0004
	SM	0.1301	0.1449	0.1410
	SP	0.6624	0.5987	0.6176
	SMP	0.0913	0.1087	0.1039
	TIME	0.0000	0.0000	0.0000
	TM	0.7768	0.6100	0.6312
	TP	0.8241	0.6501	0.6729
	TMP	0.2445	0.2646	0.2645
	ST	0.0939	0.1835	0.1752
	STM	0.2303	0.2786	0.2764
	STP	0.9455	0.7543	0.7831
	STMP	0.2149	0.2693	0.2664

ERROR TERM   EPSILON FACTORS FOR DEGREES OF FREEDOM ADJUSTMENT

	GREENHOUSE-GEISSER	HUYNH-FELDT
2	0.7292	0.8009
3	0.4023	0.4516
4	0.3041	0.3527

----

[3] Each block in the anova table requires its own correction factor when sphericity adjusted analyses are conducted. The results are interestingly consistent between the three analyses.

10.5 Example 4 – SPSS^x program MANOVA

Program 4:

SPSS-X program MANOVA analysis of the data analysed by program 1. See Table 10.1.

```
      SET WIDTH=80
[1]   DATA LIST FILE=DAT  / COND 2 PREF 3 V1 TO V12 4-63
[2]   DO REPEAT X=V1 TO V12
      COMPUTE X=LN(X-40)
      END REPEAT
[3]   MANOVA V1 TO V12 BY COND(1,2) PREF(1,2)/
[4]           WSFACTOR=EPOCH(3) TIME(4)/
              WSDESIGN=EPOCH, TIME, EPOCH BY TIME/
              PRINT=HOMOGENEITY(BOXM)/
[5]           ANALYSIS(REPEATED)/
              DESIGN/
      FINISH
```

----

[1] The data consists of 14 variables.

[2] Transform the dependent variables to remove skewness.

[3] The MANOVA routine does multivariate analysis of variance. Here variables V1 to V12, having been transformed, will be the dependent variables, and COND and PREF will be factors defining a two by two cross-classification.

[4] The twelve measures taken on each person fall into a three (EPOCH) by four (TIME) factorial structure.

[5] The REPEATED command here splits the analysis into separate overall mean, EPOCH, TIME, and EPOCH by TIME interaction analyses.

Output from Program 4:

```
   1   0              SET WIDTH=80
   2 DATA LIST FILE=DAT  / COND 2 PREF 3 V1 TO V12 4-63
THE ABOVE DATA LIST STATEMENT WILL READ    1 RECORDS FROM FILE DAT
             VARIABLE   REC   START    END         FORMAT  WIDTH  DEC

             COND        1      2       2           F        1     0
             PREF        1      3       3           F        1     0
             V1          1      4       8           F        5     0
             V2          1      9      13           F        5     0
             V3          1     14      18           F        5     0
             V4          1     19      23           F        5     0
             V5          1     24      28           F        5     0
             V6          1     29      33           F        5     0
             V7          1     34      38           F        5     0
             V8          1     39      43           F        5     0
             V9          1     44      48           F        5     0
             V10         1     49      53           F        5     0
             V11         1     54      58           F        5     0
             V12         1     59      63           F        5     0
END OF DATALIST TABLE.

   3 DO REPEAT X=V1 TO V12
   4 COMPUTE X=LN(X-40)
   5 END REPEAT
   6 MANOVA V1 TO V12 BY COND(1,2) PREF(1,2)/
   7        WSFACTOR=EPOCH(3) TIME(4)/
   8        WSDESIGN=EPOCH, TIME, EPOCH BY TIME/
   9        PRINT=HOMOGENEITY(BOXM)/
  10        ANALYSIS(REPEATED)/
  11        DESIGN/
THERE ARE   144888 BYTES OF MEMORY AVAILABLE.
THE LARGEST CONTIGUOUS AREA HAS    144888 BYTES.
>NOTE      12299
>MANOVA has changed the default type of sums of squares to UNIQUE and no longer
>prints parameter estimates by default.  To obtain SEQUENTIAL sums of squares,
>specify the subcommand / METHOD = SSTYPE(SEQUENTIAL) .   To print parameter
>estimates, specify the subcommand / PRINT = PARAMETERS(ESTIM) .


03 FEB 88 SPSS-X RELEASE 2.1  FOR IBM OS & MVS                          PAGE
15:13:02  Univ. of Lond. Computer Centre  Amdahl 5890/300     MVS/SP1.3


* * * * * * * * ANALYSIS   OF   VARIANCE * * * * * * * *

          60 cases accepted.
           0 cases rejected because of out-of-range factor values.
           0 cases rejected because of missing data.
           4 non-empty cells.
           1 design will be processed.
```

- -

[1]
Cell Number .. 1

```
Determinant of variance-covariance matrix =              .00000
LOG(Determinant) =                                    -78.87536
```

- - - - - - - - -

Cell Number .. 2

The variance-covariance matrix for this cell is singular.

- - - - - - - - -

Cell Number .. 3

```
Determinant of variance-covariance matrix =              .00000
LOG(Determinant) =                                    -79.47890
```

- - - - - - - - -

Cell Number .. 4

```
Determinant of variance-covariance matrix =              .00000
LOG(Determinant) =                                    -86.56034
```

- - - - - - - - -

```
        0 Cells with only one observation.
        1 Cell with singular variance-covariance matrix.

Determinant of pooled variance-covariance matrix        .00000
LOG(Determinant) =                                    -74.17307
```

- -

[2]
Multivariate test for Homogeneity of Dispersion matrices

```
Boxs M =                        322.88406
F with (156,4593) DF =            1.24014, P =   .024 (Approx.)
Chi-Square with 156 DF =        204.42736, P =   .006 (Approx.)
```

----

[1] The determinant of the covariance matrix is used in a test for
equality of covariance matrices, as explained in Chapter 3.

[2] There is evidence that the covariance matrices in the groups are not
identical.

- -

Correspondence between Effects and Columns of WITHIN-Subjects Design 1

```
Starting   Ending
 Column    Column    Effect Name

    1         1      CONSTANT
    2         3      EPOCH
    4         6      TIME
    7        12      EPOCH BY TIME
```

- -

```
03 FEB 88 SPSS-X RELEASE 2.1  FOR IBM OS & MVS                    PAGE   3
15:13:03  Univ. of Lond. Computer Centre  Amdahl 5890/300    MVS/SP1.3
```

* * * * * * * * * * A N A L Y S I S   O F   V A R I A N C E * * * * * * * * * *

Correspondence between Effects and Columns of BETWEEN-Subjects Design 1

```
Starting   Ending
 Column    Column    Effect Name

    1         1      CONSTANT
    2         2      COND
    3         3      PREF
    4         4      COND BY PREF
```

- -

```
03 FEB 88 SPSS-X RELEASE 2.1  FOR IBM OS & MVS                    PAGE
15:13:03  Univ. of Lond. Computer Centre  Amdahl 5890/300    MVS/SP1.3
```

* * * * * * * * * * A N A L Y S I S   O F   V A R I A N C E * * * * * * * * * *

[3]
Order of Variables for Analysis

```
 Variates      Covariates      Not Used

   *V1                            V2
                                  V3
                                  V4
```

----

[3] The analysis is being split up into four subsections, as explained in
the legend to Program 4. The first subsection is the overall mean of the
twelve scores. That is, *V1 is the overall mean and it will be analysed
alone (as a univariate analysis) initially.

```
                    V5
                    V6
                    V7
                    V8
                    V9
                    V10
                    V11
                    V12
```

 1 Dependent Variable
 0 Covariates
11 Variables not used

- -

    Note.. "*" marks TRANSFORMED variables.

            These TRANSFORMED variables correspond to the
            'CONSTANT' WITHIN-SUBJECT effect.

- -

 FEB 88 SPSS-X RELEASE 2.1 FOR IBM OS & MVS   PAGE   5
:13:03 Univ. of Lond. Computer Centre Amdahl 5890/300   MVS/SP1.3

* * * * * * * * A N A L Y S I S   O F   V A R I A N C E * * * * * * * * * *

]
ests of Significance for V1 using UNIQUE Sums of Squares

ource of Variation   SS   DF   MS   F Sig of F

ITHIN CELLS   12.371   56   .221
ONSTANT   52246.258   1 52246.258236500.582   .000
OND   .167   1   .167   .757   .388
REF   .236   1   .236   1.068   .306
OND BY PREF   .751   1   .751   3.401   .070

- -

 FEB 88 SPSS-X RELEASE 2.1 FOR IBM OS & MVS   PAGE   6
:13:03 Univ. of Lond. Computer Centre Amdahl 5890/300   MVS/SP1.3

 Univariate analysis of variance for the new V1 variable - the overall
an of the original twelve variables. This is a two way anova.
```
```

* * * * * * * * * A N A L Y S I S   O F   V A R I A N C E * * * * * * * * * *

[5]
Order of Variables for Analysis

Variates	Covariates	Not Used
*V2		V1
*V3		V4
		V5
		V6
		V7
		V8
		V9
		V10
		V11
		V12

    2 Dependent Variables
    0 Covariates
   10 Variables not used

- -

    Note.. "*" marks TRANSFORMED variables.

            These TRANSFORMED variables correspond to the
            'EPOCH' WITHIN-SUBJECT effect.

- -

[6]
Statistics for WITHIN CELLS correlations

Determinant =   .99945
Bartlett test of sphericity =   .02981 with 1 D. F.
Significance =   .863

F(max) criterion =   2.21267 with (2,56) D. F.

- -

03 FEB 88 SPSS-X RELEASE 2.1   FOR IBM OS & MVS   PAGE
15:13:04   Univ. of Lond. Computer Centre   Amdahl 5890/300   MVS/SP1.3

----

[5] *V2 and *V3 are two new variables, defined as contrasts of the
original ones, which summarise differences between EPOCHs.

[6] Sphericity test for the two EPOCH variables.

* * * * * * * * *A N A L Y S I S   O F   V A R I A N C E* * * * * * * * *

[7]
EFFECT .. COND BY PREF BY EPOCH

Multivariate Tests of Significance (S = 1, M = 0, N = 26 1/2)

Test Name	Value	Approx. F	Hypoth. DF	Error DF	Sig. of F
Pillais	.06554	1.92887	2.00	55.00	.155
Hotellings	.07014	1.92887	2.00	55.00	.155
Wilks	.93446	1.92887	2.00	55.00	.155

Roys largest root criterion =   .06554

- -

Eigenvalues and Canonical Correlations

Root No.	Eigenvalue	Pct.	Cum. Pct.	Canon. Cor.
1	.070	100.000	100.000	.256

- -

Dimension Reduction Analysis

Roots	Wilks Lambda	F	Hypoth. DF	Error DF	Sig. of F
1 TO 1	.93446	1.92887	2.00	55.00	.155

- -

[8]
Univariate F-tests with (1,56) D. F.

Variable	Hypoth. SS	Error SS	Hypoth. MS	Error MS	F
V2	.00401	.14234	.00401	.00254	1.57605
V3	.00260	.06433	.00260	.00115	2.26139

Variable	Sig. of F
V2	.215
V3	.138

- -

[7] Bivariate tests using the two EPOCH variables for the interaction aspect of the differences between groups - ie. the COND by PREF interaction.

[8] Individual tests on each of the two EPOCH derived variables.

```
03 FEB 88 SPSS-X RELEASE 2.1  FOR IBM OS & MVS                        PAGE   8
15:13:04  Univ. of Lond. Computer Centre  Amdahl 5890/300     MVS/SP1.3
```

```
* * * * * * * * * * A N A L Y S I S   O F   V A R I A N C E * * * * * * * * * *
```

[9]
EFFECT .. PREF BY EPOCH

Multivariate Tests of Significance (S = 1, M = 0, N = 26 1/2)

Test Name	Value	Approx. F	Hypoth. DF	Error DF	Sig. of F
Pillais	.05172	1.49978	2.00	55.00	.232
Hotellings	.05454	1.49978	2.00	55.00	.232
Wilks	.94828	1.49978	2.00	55.00	.232

Roys largest root criterion =   .05172

- -

Eigenvalues and Canonical Correlations

Root No.	Eigenvalue	Pct.	Cum. Pct.	Canon. Cor.
1	.055	100.000	100.000	.227

- -

Dimension Reduction Analysis

Roots	Wilks Lambda	F	Hypoth. DF	Error DF	Sig. of F
1 TO 1	.94828	1.49978	2.00	55.00	.232

- -

Univariate F-tests with (1,56) D. F.

Variable	Hypoth. SS	Error SS	Hypoth. MS	Error MS	F
V2	.00365	.14234	.00365	.00254	1.43567
V3	.00194	.06433	.00194	.00115	1.68961

Variable	Sig. of F
V2	.236
V3	.199

----

[9] Bivariate tests using the two EPOCH variables for the PREF main effect
aspect of the differences between groups.

03 FEB 88 SPSS-X RELEASE 2.1 FOR IBM OS & MVS   PAGE   9
15:13:04 Univ. of Lond. Computer Centre Amdahl 5890/300   MVS/SP1.3

* * * * * * * * * A N A L Y S I S   O F   V A R I A N C E * * * * * * * * * *

[10]
EFFECT .. COND BY EPOCH

Multivariate Tests of Significance (S = 1, M = 0, N = 26 1/2)

Test Name	Value	Approx. F	Hypoth. DF	Error DF	Sig. of F
Pillais	.17068	5.65985	2.00	55.00	.006
Hotellings	.20581	5.65985	2.00	55.00	.006
Wilks	.82932	5.65985	2.00	55.00	.006

Roys largest root criterion =   .17068

- -

Eigenvalues and Canonical Correlations

Root No.	Eigenvalue	Pct.	Cum. Pct.	Canon. Cor.
1	.206	100.000	100.000	.413

- -

Dimension Reduction Analysis

Roots	Wilks Lambda	F	Hypoth. DF	Error DF	Sig. of F
1 TO 1	.82932	5.65985	2.00	55.00	.006

- -

Univariate F-tests with (1,56) D. F.

Variable	Hypoth. SS	Error SS	Hypoth. MS	Error MS	F
V2	.02090	.14234	.02090	.00254	8.22095
V3	.00352	.06433	.00352	.00115	3.06356

Variable	Sig. of F
V2	.006
V3	.086

10] Bivariate tests using the two EPOCH variables for the COND main
ffect aspect of the differences between groups.

* * * * * * * * * A N A L Y S I S   O F   V A R I A N C E * * * * * * * * * *

[11]
 EFFECT .. EPOCH

Multivariate Tests of Significance (S = 1, M = 0, N = 26 1/2)

Test Name	Value	Approx. F	Hypoth. DF	Error DF	Sig. of F
Pillais	.00978	.27153	2.00	55.00	.763
Hotellings	.00987	.27153	2.00	55.00	.763
Wilks	.99022	.27153	2.00	55.00	.763

Roys largest root criterion =   .00978

- -

Eigenvalues and Canonical Correlations

Root No.	Eigenvalue	Pct.	Cum. Pct.	Canon. Cor.
1	.010	100.000	100.000	.099

- -

Dimension Reduction Analysis

Roots	Wilks Lambda	F	Hypoth. DF	Error DF	Sig. of F
1 TO 1	.99022	.27153	2.00	55.00	.763

- -

Univariate F-tests with (1,56) D. F.

Variable	Hypoth. SS	Error SS	Hypoth. MS	Error MS	F
V2	.00059	.14234	.00059	.00254	.23276
V3	.00035	.06433	.00035	.00115	.30737

Variable	Sig. of F
V2	.631
V3	.582

----

[11] Tests to compare the vector of overall means of the two EPOCH
variables with a vector of zeros.

- -

03 FEB 88 SPSS-X RELEASE 2.1 FOR IBM OS & MVS   PAGE 11
15:13:04 Univ. of Lond. Computer Centre Amdahl 5890/300   MVS/SP1.3

* * * * * * * * * A N A L Y S I S   O F   V A R I A N C E * * * * * * * * *

[12]
Order of Variables for Analysis

    Variates   Covariates   Not Used

    *V4   V1
    *V5   V2
    *V6   V3
                                  V7
                                  V8
                                  V9
                                  V10
                                  V11
                                  V12

    3 Dependent Variables
    0 Covariates
    9 Variables not used

- -

     Note.. "*" marks TRANSFORMED variables.

            These TRANSFORMED variables correspond to the
            'TIME' WITHIN-SUBJECT effect.

- -

Statistics for WITHIN CELLS correlations

Determinant =   .92960
Bartlett test of sphericity =   3.95411 with 3 D. F.
Significance =   .266

F(max) criterion =   1.25196 with (3,56) D. F.

- -

12] Three new variables are defined, in terms of contrasts on the
riginal ones, which describe patterns of change over TIME.

```
03 FEB 88 SPSS-X RELEASE 2.1  FOR IBM OS & MVS                    PAGE  12
15:13:05  Univ. of Lond. Computer Centre  Amdahl 5890/300    MVS/SP1.3

* * * * * * * * * A N A L Y S I S   O F   V A R I A N C E * * * * * * * * *

[13]
EFFECT .. COND BY PREF BY TIME

Multivariate Tests of Significance (S = 1, M = 1/2, N = 26 )

Test Name          Value    Approx. F  Hypoth. DF   Error DF   Sig. of F

Pillais           .06612    1.27433       3.00        54.00       .292
Hotellings        .07080    1.27433       3.00        54.00       .292
Wilks             .93388    1.27433       3.00        54.00       .292

Roys largest root criterion =          .06612

- - - - - - - - - - - - - - - - - - - - - - - - - - - - - - - - - - - - - -


Eigenvalues and Canonical Correlations

Root No.   Eigenvalue      Pct.    Cum. Pct. Canon. Cor.

   1           .071      100.000    100.000      .257

- - - - - - - - - - - - - - - - - - - - - - - - - - - - - - - - - - - - - -


Dimension Reduction Analysis

Roots    Wilks Lambda        F  Hypoth. DF    Error DF   Sig. of F

1 TO 1       .93388    1.27433      3.00        54.00        .292

- - - - - - - - - - - - - - - - - - - - - - - - - - - - - - - - - - - - - -


Univariate F-tests with (1,56) D. F.

Variable   Hypoth. SS   Error SS  Hypoth. MS   Error MS          F

V4           .00525      .25300     .00525      .00452     1.16294
V5           .00408      .20209     .00408      .00361     1.13022
V6           .00407      .23676     .00407      .00423      .96163
```

---

[13] For the three TIME variables together, multivariate tests are
conducted for the COND by PREF interaction, the PREF main effect, the COND
main effect, and to compare the vector of three means with a vector of
three zeros.

Variable	Sig. of F
V4	.285
V5	.292
V6	.331

- -

03 FEB 88 SPSS-X RELEASE 2.1 FOR IBM OS & MVS   PAGE 13
15:13:05 Univ. of Lond. Computer Centre Amdahl 5890/300   MVS/SP1.3

* * * * * * * * * A N A L Y S I S   O F   V A R I A N C E * * * * * * * * *

EFFECT .. PREF BY TIME

Multivariate Tests of Significance (S = 1, M = 1/2, N = 26)

Test Name	Value	Approx. F	Hypoth. DF	Error DF	Sig. of F
Pillais	.03858	.72233	3.00	54.00	.543
Hotellings	.04013	.72233	3.00	54.00	.543
Wilks	.96142	.72233	3.00	54.00	.543

Roys largest root criterion =   .03858

- -

Eigenvalues and Canonical Correlations

Root No.	Eigenvalue	Pct.	Cum. Pct.	Canon. Cor.
1	.040	100.000	100.000	.196

- -

Dimension Reduction Analysis

Roots	Wilks Lambda	F	Hypoth. DF	Error DF	Sig. of F
TO 1	.96142	.72233	3.00	54.00	.543

- -

Univariate F-tests with (1,56) D. F.

Variable	Hypoth. SS	Error SS	Hypoth. MS	Error MS	F
4	.00053	.25300	.00053	.00452	.11719
5	.00460	.20209	.00460	.00361	1.27570
6	.00512	.23676	.00512	.00423	1.21052

Variable	Sig. of F
V4	.733
V5	.264
V6	.276

- -

03 FEB 88 SPSS-X RELEASE 2.1 FOR IBM OS & MVS   PAGE 14
15:13:05 Univ. of Lond. Computer Centre Amdahl 5890/300   MVS/SP1.3

* * * * * * * * * A N A L Y S I S   O F   V A R I A N C E * * * * * * * * *

EFFECT .. COND BY TIME

Multivariate Tests of Significance (S = 1, M = 1/2, N = 26)

Test Name	Value	Approx. F	Hypoth. DF	Error DF	Sig. of F
Pillais	.02615	.48333	3.00	54.00	.695
Hotellings	.02685	.48333	3.00	54.00	.695
Wilks	.97385	.48333	3.00	54.00	.695

Roys largest root criterion =   .02615

- -

Eigenvalues and Canonical Correlations

Root No.	Eigenvalue	Pct.	Cum. Pct.	Canon. Cor.
1	.027	100.000	100.000	.162

- -

Dimension Reduction Analysis

Roots	Wilks Lambda	F	Hypoth. DF	Error DF	Sig. of F
1 TO 1	.97385	.48333	3.00	54.00	.695

- -

Univariate F-tests with (1,56) D. F.

Variable	Hypoth. SS	Error SS	Hypoth. MS	Error MS	F
V4	.00519	.25300	.00519	.00452	1.14821
V5	.00027	.20209	.00027	.00361	.07388
V6	.00193	.23676	.00193	.00423	.45662

Variable	Sig. of F
V4	.289
V5	.787
V6	.502

- -

03 FEB 88 SPSS-X RELEASE 2.1 FOR IBM OS & MVS   PAGE 15
15:13:05 Univ. of Lond. Computer Centre Amdahl 5890/300   MVS/SP1.3

* * * * * * * * * A N A L Y S I S   O F   V A R I A N C E * * * * * * * * * *

EFFECT .. TIME

Multivariate Tests of Significance (S = 1, M = 1/2, N = 26)

Test Name	Value	Approx. F	Hypoth. DF	Error DF	Sig. of F
Pillais	.09565	1.90383	3.00	54.00	.140
Hotellings	.10577	1.90383	3.00	54.00	.140
Wilks	.90435	1.90383	3.00	54.00	.140

Roys largest root criterion =   .09565

- -

Eigenvalues and Canonical Correlations

Root No.	Eigenvalue	Pct.	Cum. Pct.	Canon. Cor.
1	.106	100.000	100.000	.309

- -

Dimension Reduction Analysis

Roots	Wilks Lambda	F	Hypoth. DF	Error DF	Sig. of F
1 TO 1	.90435	1.90383	3.00	54.00	.140

- -

Univariate F-tests with (1,56) D. F.

Variable	Hypoth. SS	Error SS	Hypoth. MS	Error MS	F
V4	.01980	.25300	.01980	.00452	4.38283
V5	.00728	.20209	.00728	.00361	2.01670
V6	.00166	.23676	.00166	.00423	.39340

Variable	Sig. of F
V4	.041
V5	.161
V6	.533

- -

03 FEB 88 SPSS-X RELEASE 2.1 FOR IBM OS & MVS   PAGE 16
15:13:05 Univ. of Lond. Computer Centre Amdahl 5890/300   MVS/SP1.3

* * * * * * * * * * A N A L Y S I S   O F   V A R I A N C E * * * * * * * * * *

[14]
Order of Variables for Analysis

Variates	Covariates	Not Used
*V7		V1
*V8		V2
*V9		V3
*V10		V4
*V11		V5
*V12		V6

  6 Dependent Variables
  0 Covariates
  6 Variables not used

- -

    Note.. "*" marks TRANSFORMED variables.

           These TRANSFORMED variables correspond to the
           'EPOCH BY TIME' WITHIN-SUBJECT effect.

- -

Statistics for WITHIN CELLS correlations

Determinant =   .63363
Bartlett test of sphericity =   24.25950 with 15 D. F.
Significance =   .061

F(max) criterion =   2.56804 with (6,56) D. F.

- -

03 FEB 88 SPSS-X RELEASE 2.1 FOR IBM OS & MVS   PAGE 1
15:13:06 Univ. of Lond. Computer Centre Amdahl 5890/300   MVS/SP1.3

----

[14] Six new variables are defined, in terms of contrasts on the original
ones, which describe the interaction between EPOCH and TIME.

```
* * * * * * * * * A N A L Y S I S   O F   V A R I A N C E * * * * * * * * *
```

[15]
EFFECT .. COND BY PREF BY EPOCH BY TIME

Multivariate Tests of Significance (S = 1, M = 2 , N = 24 1/2)

Test Name	Value	Approx. F	Hypoth. DF	Error DF	Sig. of F
Pillais	.21051	2.26647	6.00	51.00	.051
Hotellings	.26664	2.26647	6.00	51.00	.051
Wilks	.78949	2.26647	6.00	51.00	.051

Roys largest root criterion =   .21051

- -

Eigenvalues and Canonical Correlations

Root No.	Eigenvalue	Pct.	Cum. Pct.	Canon. Cor.
1	.267	100.000	100.000	.459

- -

Dimension Reduction Analysis

Roots	Wilks Lambda	F	Hypoth. DF	Error DF	Sig. of F
1 TO 1	.78949	2.26647	6.00	51.00	.051

- -

----

[15] For the six EPOCH by TIME interaction variables, multivariate tests
for the COND by PREF interaction, the PREF main effect, the COND main
effect, and an overall comparison with the zero vector are conducted.

Univariate F-tests with (1,56) D. F.

Variable	Hypoth. SS	Error SS	Hypoth. MS	Error MS	F
V7	.00109	.11470	.00109	.00205	.52988
V8	.00016	.07332	.00016	.00131	.12527
V9	.00306	.05214	.00306	.00093	3.28763
V10	.00001	.06668	.00001	.00119	.00559
V11	.00038	.04466	.00038	.00080	.47857
V12	.00805	.05618	.00805	.00100	8.02687

Variable	Sig. of F
V7	.470
V8	.725
V9	.075
V10	.941
V11	.492
V12	.006

- -

03 FEB 88 SPSS-X RELEASE 2.1 FOR IBM OS & MVS   PAGE 18
15:13:06   Univ. of Lond. Computer Centre Amdahl 5890/300   MVS/SP1.3

* * * * * * * * * A N A L Y S I S   O F   V A R I A N C E * * * * * * * * * *

EFFECT .. PREF BY EPOCH BY TIME

Multivariate Tests of Significance (S = 1, M = 2 , N = 24 1/2)

Test Name	Value	Approx. F	Hypoth. DF	Error DF	Sig. of F
Pillais	.18813	1.96961	6.00	51.00	.087
Hotellings	.23172	1.96961	6.00	51.00	.087
Wilks	.81187	1.96961	6.00	51.00	.087

Roys largest root criterion =   .18813

- -

Eigenvalues and Canonical Correlations

Root No.	Eigenvalue	Pct.	Cum. Pct.	Canon. Cor.
1	.232	100.000	100.000	.434

- -

Dimension Reduction Analysis

Roots	Wilks Lambda	F	Hypoth. DF	Error DF	Sig. of F

1 TO 1   .81187   1.96961   6.00   51.00   .087

- -

Univariate F-tests with (1,56) D. F.

Variable	Hypoth. SS	Error SS	Hypoth. MS	Error MS	F
V7	.00326	.11470	.00326	.00205	1.59231
V8	.00014	.07332	.00014	.00131	.10788
V9	.00034	.05214	.00034	.00093	.36439
V10	.00026	.06668	.00026	.00119	.21497
V11	.00005	.04466	.00005	.00080	.05947
V12	.01169	.05618	.01169	.00100	11.64996

Variable	Sig. of F
V7	.212
V8	.744
V9	.549
V10	.645
V11	.808
V12	.001

- -

03 FEB 88 SPSS-X RELEASE 2.1 FOR IBM OS & MVS   PAGE 19
15:13:06 Univ. of Lond. Computer Centre Amdahl 5890/300   MVS/SP1.3

* * * * * * * * * A N A L Y S I S   O F   V A R I A N C E * * * * * * * * *

EFFECT .. COND BY EPOCH BY TIME

Multivariate Tests of Significance (S = 1, M = 2 , N = 24 1/2)

Test Name	Value	Approx. F	Hypoth. DF	Error DF	Sig. of F
Pillais	.03058	.26817	6.00	51.00	.949
Hotellings	.03155	.26817	6.00	51.00	.949
Wilks	.96942	.26817	6.00	51.00	.949

Roys largest root criterion =   .03058

- -

Eigenvalues and Canonical Correlations

Root No.	Eigenvalue	Pct.	Cum. Pct.	Canon. Cor.
1	.032	100.000	100.000	.175

- -

Dimension Reduction Analysis

Roots	Wilks Lambda	F	Hypoth. DF	Error DF	Sig. of F
1 TO 1	.96942	.26817	6.00	51.00	.949

- -

Univariate F-tests with (1,56) D. F.

Variable	Hypoth. SS	Error SS	Hypoth. MS	Error MS	F
V7	.00018	.11470	.00018	.00205	.08748
V8	.00001	.07332	.00001	.00131	.00786
V9	.00010	.05214	.00010	.00093	.10369
V10	.00001	.06668	.00001	.00119	.00463
V11	.00088	.04466	.00088	.00080	1.10048
V12	.00021	.05618	.00021	.00100	.21061

Variable	Sig. of F
V7	.768
V8	.930
V9	.749
V10	.946
V11	.299
V12	.648

- -

03 FEB 88 SPSS-X RELEASE 2.1 FOR IBM OS & MVS   PAGE 20
15:13:06 Univ. of Lond. Computer Centre Amdahl 5890/300   MVS/SP1.3

* * * * * * * * * A N A L Y S I S   O F   V A R I A N C E * * * * * * * * *

EFFECT .. EPOCH BY TIME

Multivariate Tests of Significance (S = 1, M = 2 , N = 24 1/2)

Test Name	Value	Approx. F	Hypoth. DF	Error DF	Sig. of F
Pillais	.16863	1.72408	6.00	51.00	.134
Hotellings	.20283	1.72408	6.00	51.00	.134
Wilks	.83137	1.72408	6.00	51.00	.134

Roys largest root criterion =   .16863

- -

Eigenvalues and Canonical Correlations

Root No.	Eigenvalue	Pct.	Cum. Pct.	Canon. Cor.
1	.203	100.000	100.000	.411

- -

Dimension Reduction Analysis

Roots	Wilks Lambda	F	Hypoth. DF	Error DF	Sig. of F
1 TO 1	.83137	1.72408	6.00	51.00	.134

- -

Univariate F-tests with (1,56) D. F.

Variable	Hypoth. SS	Error SS	Hypoth. MS	Error MS	F
V7	.00598	.11470	.00598	.00205	2.91759
V8	.00008	.07332	.00008	.00131	.06221
V9	.00545	.05214	.00545	.00093	5.84847
V10	.00141	.06668	.00141	.00119	1.18239
V11	.00024	.04466	.00024	.00080	.30675
V12	.00018	.05618	.00018	.00100	.18298

Variable	Sig. of F
V7	.093
V8	.804
V9	.019
V10	.282
V11	.582
V12	.670

10.6 Example 5 – SPSS^x program MANOVA

Program 5:

An SPSS-X MANOVA analysis of the visual acuity data introduced in Chapter 3.

```
      SET WIDTH=80
[1]   DATA LIST FILE=DAT   / L6 1-3 L18 L36 L60 R6 R18 R36 R60  4-31
      MANOVA L6 TO R60/
[2]           WSFACTOR=EYE(2) STREN(4)/
              WSDESIGN=EYE, STREN, EYE BY STREN/
              PRINT=HOMOGENEITY(BOXM)/
[3]           ANALYSIS(REPEATED)/
              DESIGN/
      FINISH
```

---

[1] There are eight variables: the response time between a stimulus and the electrical response at the back of the cortex for each eye through lenses of four different powers.

[2] Thus the eight variables are arranged in a two by four factorial structure.

[3] The REPEATED keyword causes the analysis to be grouped into subsets of variables - an overall main effect, the main effects of EYE (left or right eye) and STREN (strength of lens), and the interaction of EYE and STREN.

Output from program 5:

```
  1   0            SET WIDTH=80
  2 DATA LIST FILE=DAT   / L6 L18 L36 L60 R6 R18 R36 R60 1-32
THE ABOVE DATA LIST STATEMENT WILL READ    1 RECORDS FROM FILE DAT
            VARIABLE    REC   START    END      FORMAT  WIDTH  DEC

            L6            1      1       4        F        4     0
            L18           1      5       8        F        4     0
            L36           1      9      12        F        4     0
            L60           1     13      16        F        4     0
            R6            1     17      20        F        4     0
            R18           1     21      24        F        4     0
            R36           1     25      28        F        4     0
            R60           1     29      32        F        4     0
END OF DATALIST TABLE.

  3 MANOVA L6 TO R60/
  4        WSFACTOR=EYE(2) STREN(4)/
  5        WSDESIGN=EYE, STREN, EYE BY STREN/
  6        PRINT=HOMOGENEITY(BOXM)/
  7        ANALYSIS(REPEATED)/
  8        DESIGN/
THERE ARE   145544 BYTES OF MEMORY AVAILABLE.
THE LARGEST CONTIGUOUS AREA HAS    145544 BYTES.
>NOTE    12299
>MANOVA has changed the default type of sums of squares to UNIQUE and no longer
>prints parameter estimates by default.  To obtain SEQUENTIAL sums of squares,
>specify the subcommand / METHOD = SSTYPE(SEQUENTIAL) .  To print parameter
>estimates, specify the subcommand / PRINT = PARAMETERS(ESTIM) .


05 FEB 88 SPSS-X RELEASE 2.1  FOR IBM OS & MVS                      PAGE   2
16:29:04  Univ. of Lond. Computer Centre  Amdahl 5890/300    MVS/SP1.3


* * * * * * * * * ANALYSIS   OF   VARIANCE * * * * * * * * * *

        7 cases accepted.
        0 cases rejected because of out-of-range factor values.
        0 cases rejected because of missing data.
        1 non-empty cells.
        1 design will be processed.

- - - - - - - - - - - - - - - - - - - - - - - - - - - - - - - - - - - -
```

[1]
Correspondence between Effects and Columns of WITHIN-Subjects Design 1

Starting Column	Ending Column	Effect Name
1	1	CONSTANT
2	2	EYE
3	5	STREN
6	8	EYE BY STREN

- -

05 FEB 88 SPSS-X RELEASE 2.1 FOR IBM OS & MVS   PAGE   3
16:29:05   Univ. of Lond. Computer Centre   Amdahl 5890/300   MVS/SP1.3
[2]
* * * * * * * * * A N A L Y S I S   O F   V A R I A N C E * * * * * * * * * *

Correspondence between Effects and Columns of BETWEEN-Subjects Design 1

Starting Column	Ending Column	Effect Name
1	1	CONSTANT

- -

05 FEB 88 SPSS-X RELEASE 2.1 FOR IBM OS & MVS   PAGE
16:29:05   Univ. of Lond. Computer Centre   Amdahl 5890/300   MVS/SP1.3

* * * * * * * * * A N A L Y S I S   O F   V A R I A N C E * * * * * * * * * *

Order of Variables for Analysis

Variates	Covariates	Not Used
*L6		L18
		L36
		L60
		R6
		R18
		R36
		R60

----

[1] The factorial structure within subjects means that one can define
combinations of the original variables which describe the overall mean of
the original variables, the difference between eyes, the differences
between the lenses, and the interaction of eye and lens strength.

[2] The analysis proceeds through each of the 'new' sets of variables in
turn, beginning with the overall average or constant.

1 Dependent Variable
0 Covariates
7 Variables not used

- -

   Note.. "*" marks TRANSFORMED variables.

        These TRANSFORMED variables correspond to the
        'CONSTANT' WITHIN-SUBJECT effect.

- -

5 FEB 88 SPSS-X RELEASE 2.1 FOR IBM OS & MVS   PAGE   5
6:29:05 Univ. of Lond. Computer Centre Amdahl 5890/300   MVS/SP1.3

3]
* * * * * * * * A N A L Y S I S   O F   V A R I A N C E * * * * * * * * *

Tests of Significance for L6 using UNIQUE Sums of Squares

Source of Variation	SS	DF	MS	F	Sig of F
WITHIN CELLS	1379.429	6	229.905		
CONSTANT	720271.446	1	720271.446	3132.912	.000

- -

5 FEB 88 SPSS-X RELEASE 2.1 FOR IBM OS & MVS   PAGE   6
6:29:06 Univ. of Lond. Computer Centre Amdahl 5890/300   MVS/SP1.3

6]
* * * * * * * * A N A L Y S I S   O F   V A R I A N C E * * * * * * * * *

Order of Variables for Analysis

Variates	Covariates	Not Used
*L18		L6
		L36
		L60
		R6
		R18
		R36
		R60

Since there is only one group of subjects, the 'analysis of variance' for
the overall average is exceptionally simple, involving just a single
test comparing a group mean with zero.

   The next analysis of variance uses the difference between the two eyes
as the dependent variable (averaged over the four lens strengths)

    1 Dependent Variable
    0 Covariates
    7 Variables not used

- -

    Note.. "*" marks TRANSFORMED variables.

        These TRANSFORMED variables correspond to the
        'EYE' WITHIN-SUBJECT effect.

- -

05 FEB 88 SPSS-X RELEASE 2.1 FOR IBM OS & MVS   PAGE   7
16:29:06 Univ. of Lond. Computer Centre Amdahl 5890/300   MVS/SP1.3

* * * * * * * * * A N A L Y S I S   O F   V A R I A N C E * * * * * * * * *

[5]
 Tests of Significance for L18 using UNIQUE Sums of Squares

 Source of Variation   SS   DF   MS   F   Sig of F

 WITHIN CELLS   357.429   6   59.571
 EYE   46.446   1   46.446   .780   .411

- -

05 FEB 88 SPSS-X RELEASE 2.1 FOR IBM OS & MVS   PAGE   8
16:29:06 Univ. of Lond. Computer Centre Amdahl 5890/300   MVS/SP1.3

[6]
* * * * * * * * * A N A L Y S I S   O F   V A R I A N C E * * * * * * * * *

 Order of Variables for Analysis

    Variates   Covariates   Not Used

    *L36   L6
    *L60   L18
    *R6   R18
                                  R36
                                  R60

----

[5] the difference between the eyes is non-significant.

[6] The next analysis of variance is a comparison of the four lens
strengths. The space of differences between the four values is spanned by
a three dimensional vector, so this analysis will be multivariate -
trivariate.

3 Dependent Variables
0 Covariates
5 Variables not used

- -

     Note.. "*" marks TRANSFORMED variables.

          These TRANSFORMED variables correspond to the
          'STREN' WITHIN-SUBJECT effect.

- -

Statistics for WITHIN CELLS correlations

Determinant =   .06465
Bartlett test of sphericity =   11.41133 with 3 D. F.
Significance =   .010

F(max) criterion =   7.02587 with (3,6) D. F.

- -

5 FEB 88 SPSS-X RELEASE 2.1 FOR IBM OS & MVS   PAGE   9
6:29:07 Univ. of Lond. Computer Centre Amdahl 5890/300   MVS/SP1.3

7]
 * * * * * * * * A N A L Y S I S   O F   V A R I A N C E * * * * * * * * *

EFFECT .. STREN

Multivariate Tests of Significance (S = 1, M = 1/2, N = 1)

Test Name   Value   Approx. F Hypoth. DF   Error DF   Sig. of F

Pillais   .94158   21.49012   3.00   4.00   .006
Hotellings   16.11759   21.49012   3.00   4.00   .006
Wilks   .05842   21.49012   3.00   4.00   .006

Roys largest root criterion =   .94158

- -

----

*] The multivariate tests show a highly significant result. There are
differences between the effects of the different lenses.

Eigenvalues and Canonical Correlations

Root No.	Eigenvalue	Pct.	Cum. Pct.	Canon. Cor.
1	16.118	100.000	100.000	.970

- -

Dimension Reduction Analysis

Roots	Wilks Lambda	F	Hypoth. DF	Error DF	Sig. of F
1 TO 1	.05842	21.49012	3.00	4.00	.006

- -

Univariate F-tests with (1,6) D. F.

Variable	Hypoth. SS	Error SS	Hypoth. MS	Error MS	F
L36	75.57143	80.42857	75.57143	13.40476	5.63766
L60	6.85714	36.80952	6.85714	6.13492	1.11772
R6	58.33929	258.61905	58.33929	43.10317	1.35348

Variable	Sig. of F
L36	.055
L60	.331
R6	.289

- -

05 FEB 88 SPSS-X RELEASE 2.1 FOR IBM OS & MVS   PAGE 1
16:29:07 Univ. of Lond. Computer Centre Amdahl 5890/300   MVS/SP1.3

[8]
* * * * * * * * * A N A L Y S I S   O F   V A R I A N C E * * * * * * * * * *

Order of Variables for Analysis

Variates	Covariates	Not Used
*R18		L6
*R36		L18
*R60		L36
		L60
		R6

----

[8] The final analysis uses the block of variables describing the
interaction between eye and lens strength.

3 Dependent Variables
0 Covariates
5 Variables not used

- -

Note.. "*" marks TRANSFORMED variables.

These TRANSFORMED variables correspond to the
'EYE BY STREN' WITHIN-SUBJECT effect.

- -

Statistics for WITHIN CELLS correlations

Determinant =   .73399
Bartlett test of sphericity =   1.28856 with 3 D. F.
Significance =   .732

F(max) criterion =   7.54927 with (3,6) D. F.

- -

05 FEB 88 SPSS-X RELEASE 2.1   FOR IBM OS & MVS   PAGE 11
16:29:08 Univ. of Lond. Computer Centre Amdahl 5890/300   MVS/SP1.3

[9]
* * * * * * * * * * A N A L Y S I S   O F   V A R I A N C E * * * * * * * * *

EFFECT .. EYE BY STREN

Multivariate Tests of Significance (S = 1, M = 1/2, N = 1)

Test Name	Value	Approx. F	Hypoth. DF	Error DF	Sig. of F
Pillais	.29290	.55231	3.00	4.00	.673
Hotellings	.41423	.55231	3.00	4.00	.673
Wilks	.70710	.55231	3.00	4.00	.673

Roys largest root criterion =   .29290

- -

----

[9] The interaction appears non-significant. That is, the differences
between the lens strengths do not differ between the eyes.

Eigenvalues and Canonical Correlations

Root No.   Eigenvalue   Pct.   Cum. Pct. Canon. Cor.

     1   .414   100.000   100.000   .541

- -

Dimension Reduction Analysis

Roots   Wilks Lambda   F Hypoth. DF   Error DF   Sig. of F

1 TO 1   .70710   .55231   3.00   4.00   .673

- -

Univariate F-tests with (1,6) D. F.

Variable   Hypoth. SS   Error SS Hypoth. MS   Error MS   F

R18   2.28571   22.71429   2.28571   3.78571   .60377
R36   1.19048   36.80952   1.19048   6.13492   .19405
R60   37.14881 171.47619   37.14881   28.57937   1.29985

Variable   Sig. of F

R18   .467
R36   .675
R60   .298

- -

    15024 BYTES OF WORKSPACE NEEDED FOR MANOVA EXECUTION.

05 FEB 88 SPSS-X RELEASE 2.1   FOR IBM OS & MVS   PAGE 1:
16:29:09 Univ. of Lond. Computer Centre   Amdahl 5890/300   MVS/SP1.3

PRECEDING TASK REQUIRED   0.10 SECONDS CPU TIME;   6.58 SECONDS ELAPSED

    9 FINISH
    9 COMMAND LINES READ.
    0 ERRORS DETECTED.
    0 WARNINGS ISSUED.
    0 SECONDS CPU TIME.
    8 SECONDS ELAPSED TIME.
      END OF JOB.

10.7 Example 6 – BMDP program 5V

Program 6:

BMDP program 5V analysis of the rat data of Table 2.4. In this
example only the measurements on days 1, 8, 15, and 22 are used.
There are three groups of rats, receiving different diets.

```
[1]  /input     file is '\rat.dat'.
[2]             variables are 12.
                format is free.
[3]  /variables use = 1 to 5.
[4]              names = group, var1, var2, var3, var4.
[5]  /design    grouping = 1.
                level = 4.
                repeated = time.
                contrast(time) = x.
                x = 1,8,15,22.
                dpname = weight.
                dpvar = 2 to 5.
[6]  /model     weight = 'group*x'.
[7]  /structure type = random.
                zcol1 = 1,1,1,1.
                zcol2 = 1,8,15,22.
     /print     linesize = 80.
     /end
```

----

[1] Specifies the name of the data file.
[2] Twelve variables are read in.
[3] Only the first five variables are used.
[4] The variables to be used are given names.
[5] The DESIGN paragraph states that the between rats grouping
    factor is the first variable, that the repeated measurements
    factor has four levels and is called 'time', and that its
    levels are given by a variable x as 1, 8, 15, and 22. The
    dependent variable is given the name 'weight', and the scores
    are to be found in input variables identified by the dpvar
    line.
[6] The model specification is a complete factorial on group by
    the newly defined variable x.
[7] The covariance matrix is taken as having a random effects
    structure with zcol1 and zcol2 giving the first and second
    column, respectively, of the Z matrix

Output from program 6:

```
NUMBER OF VARIABLES TO READ IN. . . . . . . . .      12
NUMBER OF VARIABLES ADDED BY TRANSFORMATIONS. .       0
TOTAL NUMBER OF VARIABLES . . . . . . . . . . .      12
CASE WEIGHT VARIABLE. . . . . . . . . . . . .
CASE LABELING VARIABLES . . . . . . . . . . .
NUMBER OF CASES TO READ IN. . . . . . . . . . . TO END
MISSING VALUES CHECKED BEFORE OR AFTER TRANS. . NEITHER
BLANKS ARE. . . . . . . . . . . . . . . . . . . MISSING
INPUT FILE. . .\sigma\cours\data6.dat
REWIND INPUT UNIT PRIOR TO READING. . DATA. . .     YES
NUMBER OF WORDS OF DYNAMIC STORAGE. . . . . .      16298
```

```
VARIABLES TO BE USED
      1 group        2 var1       3 var2        4 var3       5 var4
PAGE   2  BMDP5V 01/13/90                   03:11:33
```

```
INPUT FORMAT IS
FREE

MAXIMUM LENGTH DATA RECORD IS   80 CHARACTERS.
ALGORITHM NAME. . . . . . . . . . . . . .NR
MAXIMUM NUMBER OF ITERATIONS. . . . . . 15
CONVERGENCE CRITERION . . . . . . . . .LOG
TOLERANCE FOR CONVERGENCE . . . . . .     0.10000E-04
```

```
DESIGN SPECIFICATIONS

          GROUP =    1
          DEPEND =   2   3   4   5
          LEVEL =    4
```

VARIABLE		STATED VALUES FOR			GROUP	CATEGORY	INTERVALS	
NO.	NAME	MINIMUM	MAXIMUM	MISSING	CODE	INDEX	NAME	.GT.	.LE.
1	group				1.000	1	*1		
					2.000	2	*2		
					3.000	3	*3		

NOTE: CATEGORY NAMES BEGINNING WITH * WERE CREATED BY THE PROGRAM.

PAGE   3   BMDP5V 01/13/90   03:11:33

GROUP STRUCTURE

```
group       COUNT
*1            8
*2            4                                    [1]
*3            4
```

      DECODED FORMULA

group + x + group.x   [2]

INTERCEPT IS INCLUDED IN THE MODEL .

ANALYSIS FOR   1ST DEPENDENT VARIABLE

Z MATRIX IN THE RANDOM EFFECTS STRUCTURE


```
                    1          2

      1          1.000      1.000              [3]
      2          1.000      8.000
      3          1.000     15.000
      4          1.000     22.000
```

```
NUMBER OF CASES READ. . . . . . . . . . . . . . . . .    16
    CASES CONTRIBUTING DATA TO THE ANALYSIS . . . . .    16
  CASES WITH COVAR. OR DEP.VAR. MISSING OR BEYOND LIMITS  0
    COMPLETE CASES (CASES WITH NO MISSING DATA) . . .    16
  TOTAL NUMBER OF MEASUREMENTS . . . . . . . . . . .     64
```

----

[1] Numbers in groups.

[2] Model to be fitted.

[3] Z matrix for the random effects.

ITER NO	STEP HLVS	LOG LIKELIHOOD	CHG IN LOG LIKELIHOOD	MAX CHG IN COV.PAR
0	0	-405.14		
1	0	-231.20	0.17394E+03	0.75764E+04
2	0	-231.20	0.60254E-11	0.15854E-07

```
MAXIMUM LOG LIKELIHOOD          =      -231.204      [4]
CHANGE SINCE LAST STEP          =     0.000000000

APPROX. CONVERGENCE RATE        =     0.000000000
MIN. TOL. FOR COV. MATRIX       =     0.018735005


PAGE   4  BMDP5V 01/13/90                    03:11:33
```

AKAIKE'S INFORMATION CRITERION(AIC):

AIC MAY BE USED TO SELECT AN APPROPRIATE COVARIANCE
STRUCTURE BY RUNNING 5V WITH DIFFERENT TYPES OF
STRUCTURES AND SELECTING THE ONE WITH MAXIMUM AIC VALUE

```
AIC                             =      -235.204
```

ESTIMATES OF COVARIANCE PARAMETERS:

WITHIN-SUBJECT COVARIANCE MATRIX IS A FUNCTION
OF COVARIANCE PARAMETERS. 5V PROVIDES SEVERAL
BUILT-IN STRUCTURES USERS CAN CHOOSE FROM BY
SPECIFYING THE TYPE IN THE STRUCTURE PARAGRAPH

----

[4] The difference between this maximised log-likelihood and
    the figure for the model with an unstructured covariance
    matrix can be used to test the adequacy of this model.

COV. STRUCTURE TYPE=RANDOM EFFECTS

```
************************************
SEE THE BMDP MANUAL FOR EXPLANATIONS
************************************
```

COV PAR	ESTIMATE	ASYMPTOTIC SE	Z	
1	1118.3564	399.5368	2.799	[5]
2	-2.5282	3.6791	-0.687	
3	0.1217	0.0661	1.841	
4	14.7813	3.6953	4.000	

MODEL: WITHIN-SUBJECT COVARIANCE MATRIX

	1	2	3	4
1	1128.2030			
2	1096.5763	1100.4755		
3	1079.7308	1074.8123	1084.6752	
4	1062.8853	1063.9304	1064.9755	1080.8018

ALL-PAIRS WITHIN-SUBJECT COVARIANCE MATRIX

	1	2	3	4
1	1161.4439			
2	1110.6980	1095.9870		
3	1045.8348	1032.9634	1004.7717	
4	1106.2030	1095.3516	1051.8597	1131.9529

PAGE   5   BMDP5V 01/13/90   03:11:33

----

[5] Estimates of the parameters in the dispersion matrix and
    variance in the random effects model.

MODEL: WITHIN-SUBJECT CORRELATION MATRIX

	1	2	3	4
1	1.0000			
2	0.9841	1.0000		
3	0.9760	0.9838	1.0000	
4	0.9625	0.9756	0.9836	1.0000

ALL-PAIRS WITHIN-SUBJECT CORRELATION MATRIX

	1	2	3	4
1	1.0000			
2	0.9845	1.0000		
3	0.9681	0.9843	1.0000	
4	0.9648	0.9834	0.9863	1.0000

ESTIMATES OF REGRESSION PARAMETERS:

IN THE TABLE BELOW EACH FIXED EFFECT PART OF THE MODEL IS
DECOMPOSED INTO SINGLE DEGREE OF FREEDOM REGRESSION TERMS
AND COVARIATES

PARAMETER		ESTIMATE	ASYMPTOTIC SE	Z	TWO-SIDED P-VALUE
1	CONST.	402.73988	8.85857	45.463	0.0000
2	g1	-152.71310	11.20531	-13.629	0.0000
3	g2	49.61726	13.13939	3.776	0.0002
4	x	0.66845	0.11243	5.945	0.0000
5	g1.x	-0.19524	0.14222	-1.373	0.1698
6	g2.x	0.34940	0.16677	2.095	0.0362

PAGE   6   BMDP5V 01/13/90   03:11:33

WALD TESTS OF SIGNIFICANCE OF
FIXED EFFECTS AND COVARIATES

TEST	DF	CHI-SQUARE	P-VALUE
g	2	190.80	0.000
x	1	35.35	0.000
g.x	2	4.67	0.097

Bibliography

This list incorporates ones prepared by Dr Ian Ford, Dr Mike Kenward and Dr Andrew Pickles. We are most grateful for permission to use their work thus. A consequence is that many of the references appearing below have not been mentioned in the text. Those that have, are distinguished by an asterisk. Also, the list is representative but nowwhere near comprehensive. This is purely because of the topics covered in this book, and their treatment, and has nothing to do with the quality or importance of work in this field. Other references, cited in the book but not explicitly about repeated measures, are given in the Supplementary list.

Abeyasekara, S. and Curnow, R.N. (1984) The desirability of adjusting for residual effects in a crossover design. *Biometrics*, **40**, 1071–1078.

Afsarinejad, K. (1983) Balanced repeated measurements designs. *Biometrika*, **70**, 199–204.

Aitkin, M. (1981) Regression models for repeated measurements. *Biometrics*, **37**, 831–832.

Allen, O.B. (1983) A guide to the analysis of growth curve data with special reference to SAS. *Computers in Biomedical Research*, **16**, 101–115.

Amara, I.A., Koch, G.G. and Markley, J.B. (1977) A statistical strategy for non-parametric analysis of longitudinal data from a two-period change-over design. *Statistical Computing Section, Proceedings of the American Statistical Association*, pp. 108–113.

Anderson, A.H., Jensen, E.B. and Schou, G. (1981) Two-way analysis of variance with correlated errors. *International Statistical Review*, **49**, 153–167.

*Anderson, T.W. (1978) Repeated measurements on autoregressive processes. *Journal of the American Statistical Association*, **73**, 371–378.

Armitage, P. and Hills, M. (1982) The two-period crossover trial. *The Statistician*, **31**, 119–131.

Azzalini, A. (1981) Replicated observations of low order autoregressive time series. *Journal of Time Series Analysis*, **2**, 63–70.

*Azzalini, A. (1984) Estimation and hypothesis testing for collections of autoregressive time series. *Biometrika*, **71**, 85–90.

*Azzalini, A. (1987) Growth curves analysis for patterned covariance matrices, in *New Perspectives in Theoretical and Applied Statistics* (eds M. Puri, J.P. Vilaplana and W. Wertz). Wiley, New York, pp. 63–73.

Azzalini, A. and Giovagnoli, A. (1987) Some optimal designs for repeated measurements with autoregressive errors. *Biometrika*, **74**, 725–734.

Baksalary, J.K., Corsten, L.C.A. and Kala, R. (1978) Reconciliation of two different views on estimation of growth curve parameters. *Biometrika*, **65**, 661–665.

Barcikowski, R.S. and Robey, R.R. (1984) Decisions in single group repeated measures analysis: statistical tests and three computer packages. *American Statistician*, **38**, 148–150.

Barker, N., Hews R.J., Huitson A., and Ploniecki, J. (1982) The two period cross over trial. *Bulletin in Applied Statistics*, **9**, 67–116.

*Beach, C.M. and Mackinnon, J.G. (1978) A maximum likelihood procedure for regression with autocorrelated errors. *Econometrica*, **46**, 51–58.

Berg, L. (1985) Numerical results on growth functions. *Biometrical Journal*, **27**, 565–580.

Berk, K. (1984) Computing for unbalanced repeated measures experiments. *Proceedings of the Eleventh Annual SUGI Conference*, Cary, North Carolina.

Berk, K. (1987) Computing for incomplete repeated measures. *Biometrics*, **43**, 385–398.

*Berkey, C.S. (1982a) Bayesian approach for a nonlinear growth model. *Biometrics*, **38**, 953–961.

*Berkey, C.S. (1982b) Comparison of two longitudinal growth models for pre-school children. *Biometrics*, **38**, 221–234.

Bhapkar, V.P. and Patterson, K.W. (1977) On some non-parametric tests for profile analyses of several multivariate samples. *Journal of Multivariate Analysis*, **7**, 265–277.

Bhapkar, V.P. and Patterson, K.W. (1978) A Monte Carlo study of some multivariate non-parametric statistics for profile analysis of several samples. *Journal of Statistical Computation and Simulation*, **6**, 223–237.

*Bishop S.H. and Jones, B. (1984) A review of higher order cross-over designs. *Journal of Applied Statistics*, **11**, 29–50.

Bjornsson, H. (1978a) Analysis of responses to long term P-fertilizer application when the errors are autoregressive. *Journal of Agricultural Research, Iceland*, **10**, 22–33.

Bjornsson, H. (1978b) Analysis of a series of long-term grassland experiments with autocorrelated errors. *Biometrics*, **34**, 645–651.

Blumen, J., Kogan, M. and McCarthy, D.J. (1955) *The Industrial Mobility of Labour as a Probability Process*. Cornell University Press, Ithaca, New York.

Bock, R.D. (1967) Multivariate analysis of variance of repeated measurements, in *Problems in Measuring Change* (ed. C.W. Harris). University of Wisconsin Press, Madison, pp. 85–103.

Bock, R.D. (1979) Univariate and multivariate analysis of variance of time-structured data. Chapter 8 in *Longitudinal Research in the Study of Behaviour and Development* (eds J.R. Nesselroade and P.B. Baltes). Academic Press, New York, pp. 199–232.

*Boick, R.J. (1981) A priori tests in repeated measures designs: Effect of nonsphericity. *Psychometrika*, **46**, 241–255.

Bora, A.C. (1984) Change-over designs with errors following a first order autoregressive process. *Australian Journal of Statistics*, **26**, 179–188.

Bora, A.C. (1985) Change-over design with first-order residual effects and errors following a first-order autoregressive process. *The Statistician*, **34**, 161–173.

Bowden, D.C. and Steinhorst, R.K. (1973) Tolerance bands for growth curves. *Biometrics*, **29**, 361–371.

Box, G.E.P. (1950) Problems in the analysis of growth and wear curves. *Biometrics*, **6**, 362–389.

*Brown, B.W.Jr. (1980) The crossover experiment for clinical trials. *Biometrics*, **36**, 69–79.

Bryant, E. and Gillings, D. (1985) Statistical analysis of longitudinal repeated measures designs, in *Biostatistics: Statistics in Biomedical Public Health and Environmental Sciences* (ed. P.K. Sen). Elsevier Science Publishers, North-Holland.

*Byrne, P.J. and Arnold, S.F. (1983) Inference about multivariate means for a nonstationary autoregressive model. *Journal of the American Statistical Association*, **78**, 850–855.

Carter, E.M. and Hubert, J.J. (1984) A growth-curve model approach to multivariate quantal bio-assay. *Biometrics*, **40**, 699–706.

*Carter R.L. and Yang M.C.K. (1986) Large sample inference in random coefficient regression models. *Communications in Statistics – Theory and Methods*, **15**, 2507–2525.

Castellana, J.V. and Patel, H.I. (1985) Analysis of two-period crossover design in a multicenter clinical trial. *Biometrics*, **41**, 969–977.

Causton, D.R. and Venus, J.C. (1981) *The Biometry of Plant Growth*. Edward Arnold, London.

Chamberlain, G. (1984) Panel data. Chapter 22 in *Handbook of Econometrics*, vol. II (eds Z. Grilliches and M.D. Intrilligator). Elsevier Science Publishers, New York, pp. 1247–1318.

Cheng, C.S. and Wu, C.F. (1980) Balanced repeated measurements designs. *Annals of Statistics*, **8**, 1272–1283.

Chinchilli, V.M. and Carter, W.H., Jr. (1984) A likelihood ratio test for a patterned covariance matrix in a multivariate growth-curve model. *Biometrics*, **40**, 151–156.

Church, A., Jr. (1966) Analysis of data when the response is a curve. *Technometrics*, **8**, 229–246.

Church, C.K. and Schwenka, J.R. (1986) Autoregressive errors with a repeated measures design in clinical trials. *Controlled Clinical Trials*, **7**, 149–164.

Clayton D. (1985) Using test–retest reliability data to improve estimates of relative risk: an application of latent class analysis. *Statistics in Medicine*, **4**, 445–455.

Clayton, D. and Cuzick, J. (1985) Multivariate generalizations of the proportional hazards model. *Journal of the Royal Statistical Society*, A, **148**, 82–117.

Cole, J.W.L. and Grizzle, J.E. (1966) Applications of multivariate analysis of variance to repeated measurements experiments. *Biometrics*, **22**, 810–828.

*Collier, R.O., Jr., Baker, F.B., Mandeville, G.K. and Hayes, T.F. (1967) Estimates of test size for several test procedures based on conventional variance ratios in the repeated measures design. *Psychometrika*, **32**, 339–353.

Cook, N.R. (1982) *A General Linear Model Approach to Longitudinal Data Analysis*. Ph.D. dissertation, Department of Biostatistics, Harvard School of Public Health.

Cook, N.R. and Ware, J.H. (1983) Design and analysis methods for longitudinal research. *Annual Review of Public Health*, **4**, 1–24.

*Crépeau, H., Koziol, J., Reid, N., and Yuh, Y.S. (1985) Analysis of incomplete multivariate data from repeated measurement experiments. *Biometrics*, **41**, 505–514.

Crowder, M.J. (1978) On concurrent regression lines. *Applied Statistics*, **27**, 310–318.

Crowder, M.J. (1980) Proportional linear models. *Applied Statistics*, **29**, 299–303.

*Crowder, M.J. (1983) A growth curve analysis for EDP curves. *Applied Statistics*, **32**, 15–18.

*Crowder, M.J. (1985) A distributional model for repeated failure time measurements. *Journal of the Royal Statistical Society*, B, **47**, 447–452.

Crowder, M.J. (1989) A multivariate distribution with Weibull connections. *Journal of the Royal Statistical Society*, B, **51**, 93–107.

*Crowder, M.J. and Tredger, J.A. (1981) The use of exponentially damped polynomials for biological recovery data. *Applied Statistics*, **30**, 147–152. (1983) Correction. *Applied Statistics*, **31**, 63.

Danford, M.B., Hughes, H.M. and McNee, R.C. (1960) On the analysis of repeated measurements experiments. *Biometrics*, **16**, 547–565.

Darby, S.C. and Fearn, T. (1979) The Chatham blood pressure study. An application of Bayesian growth curve models to a longitudinal study of blood pressure in children. *International Journal of Epidemiology*, **8**, 15–21 and 157–175.

Davidson, J. and Turnbull, C.D. (1982) Loss of appetite and weight associated with isocarboxazid in depression. *Journal of Clinical Psychopharmacology*, **2**, 263–266.

Davidson, J., Weiss, J., Sullivan, J., Turnbull, C.D. and Linnoila, M. (1981) A placebo-controlled evaluation of isocarboxazid in outpatients, in *Monamine Oxidase Inhibitors: The State of the Art* (eds Youdin and Paykel). Wiley, New York, pp. 115–123.

Davidson, M.L. (1972) Univariate versus multivariate tests in repeated-measures experiments. *Psychological Bulletin*, **77**, 446–452.

Davidson, M.L. (1975) The multivariate approach to repeated measures. *BMDP Statistical Software*. Technical Report No. 75., Dept. of Biomathematics, University of California, Los Angeles.

Day, N.E. (1966) Fitting curves to longitudinal data. *Biometrics*, **22**, 276–291.

*Dempster, A.P., Rubin, D.B. and Tsutakawa, R.K. (1981) Estimation in covariance component models. *Journal of the American Statistical Association*, **76**, 341–353.

Dempster, A.P., Selwyn, M.R., Patel, C.M. and Roth, A.J. (1984) Statistical and computational aspects of mixed model analysis. *Applied Statistics*, **33**, 203–214.

*Diggle, P.J. (1988) An approach to the analysis of repeated measures. *Biometrics*, **44**, 959–971.

Dunsmore, I.R. (1981a) The analysis of preferences in crossover designs. *Biometrics*, **37**, 575–578.

*Dunsmore, I.R. (1981b) Growth curves in two-period change over models. *Applied Statistics*, **30**, 223–229.

Elashoff, J.D. (1981) Repeated measures bioassay with correlated errors and heterogeneous variances: A Monte Carlo study. *Biometrics*, **37**, 475–482.

Elston, R.C. and Grizzle, J.E. (1962) Estimation of time-response curves and their confidence bands. *Biometrics*, **18**, 148–159.

Estevè, J. and Schifflers, E. (1974) Some aspects of growth in laboratory mice (statistical analysis of growth curves). *Annals of Zoology – Ecology Animal*, **6**, 449–462.

Estevè, J. and Schifflers, E. (1976) Discussion et illustration de quelques methodes d'analyse longitudinale. *Proceedings of the 9th International Biometric Conference*, **1**, 463–479.

Evans, J.C. and Roberts, E.A. (1979) Analysis of sequential observations with applications to experiments on grazing animals and perennial plants. *Biometrics*, **35**, 687–694.

*Farewell, V.T. (1985) Some remarks on the analysis of cross-over trials with a binary response. *Applied Statistics*, **34**, 121–128.

*Fearn, T. (1975) A Bayesian approach to growth curves. *Biometrika*, **62**, 89–100.

*Fearn, T. (1977) A two-stage model for growth curves which leads to Rao's covariance adjusted estimators. *Biometrika*, **64**, 141–143.

Federer, W.T. (1975) The misunderstood split plot, in *Applied Statistics* (ed. R.P. Gupta). North Holland Publishing Co., Amsterdam, pp. 9–39.

*Fidler, V. (1984) Change-over clinical trial with binary data: mixed-model based comparison of tests. *Biometrics*, **40**, 1063–1070.

Fidler, V. (1988) A log-linear model for binary cross-over data. *Applied Statistics*, **37**, 111–113.

Finney, D.J. (1982) Intervention and correlated sequences of observations. *Biometrics*, **38**, 255–267.

*Fisk, P.R. (1967) Models of the second kind in regression analysis. *Journal of the Royal Statistical Society*, B, **29**, 266–281.

*Foulkes, M.A. and Davis, C.E. (1981) An index of tracking for longitudinal data. *Biometrics*, **37**, 439–446.

*Foutz, R.V., Jensen, D.R. and Anderson, G.W. (1985) Multiple comparisons in the randomisation analysis of designed experiments with growth curve responses. *Biometrics*, **41**, 29–37.

*France, J. and Dhanova, M.S. (1984) On estimating location yield. *Journal of Agricultural Science, Cambridge*, **103**, 245–247.

Frane, J.W. (1980) The univariate approach to repeated measures. *BMDP Statistical Software*. Technical Report No. 69, Dept. of Biomathematics, University of California, Los Angeles.

*Freeman, P.R. (1987) The performance of the Hills Armitage analysis of two-treatment crossover trials. (Manuscript).

*Gabriel, K.R. (1961) The model of ante-dependence for data of biological growth. *Bulletin Institut 33rd Session* (Paris), 253–264.

Gabriel, K.R. (1962) Ante-dependence analysis of an ordered set of variables. *Annals of Mathematical Statistics*, **33**, 201–212.

*Gart, J.J. (1969) An exact test for comparing matched proportions in crossover designs. *Biometrika*, **56**, 75–80.

Gasser, T., Muller, H-G. and Kohler, W. (1984) Non-parametric regression analysis of growth curves. *Annals of Statistics*, **12**, 210–229.

*Geary, D.N. (1988) Sequential testing in clinical trials with repeated measurements. *Biometrika*, **75**, 311–318.

Geisser, S. and Greenhouse, S.W. (1958) An extension of Box's results on the use of the *F*-distribution in multivariate analysis. *Annals of Mathematical Statistics*, **29**, 885–891.

Geisser, S. (1963) Multivariate analysis of variance for a special covariance case. *Journal of the American Statistical Association*, **58**, 660–669.

Geisser, S. (1970) Bayesian analysis of growth curves. *Sankhya A*, **32**, 53–64.

Ghosh, M., Grizzle, J.E. and Sen, P.K. (1973) Nonparametric methods in longitudinal studies. *Journal of the American Statistical Association*, **68**, 29–36.

Gill, J.L. and Hafs, H.D. (1971) Analysis of repeated measurements of animals. *Journal of Animal Science*, **35**, 181–189.

Gill, J.L. (1979) combined significance of non-independent tests for repeated measurements. *Journal of Animal Science*, **48**, 363–366.

*Gill, R.D. (1986) On estimating transition intensities of a Markov process with aggregate data of a certain type: 'occurrences but no exposures'. *Scandinavian Journal of Statistics*, **13**, 113–134.

Gillings, D.B., Symons, M.J. and Donelan, M.M. (1978) Strategies for the analysis of repeated measurements of designs involving ordinal categorical data. *Institute of Statistics Mimeo Series No. 1215*. Department of Biostatistics, University of North Carolina, Chapel Hill.

Glasbey, C.A. (1979) Correlated residuals in non-linear regression applied to growth data. *Applied Statistics*, **28**, 251–259.

Glasbey, C.A. (1980) Nonlinear regression with autoregressive time-series errors. *Biometrics*, **36**, 139–140.

Goldstein, H. (1968) Longitudinal studies and measurement of change. *The Statistician*, **18**, 93–117.

*Goldstein, H. (1979a) *The Design and Analysis of Longitudinal Studies*. Academic Press, New York.

*Goldstein, H. (1979b) Some models for analysing longitudinal data on educational attainment (with discussion). *Journal of the Royal Statistical Society*, A, **142**, 407–442.

*Goldstein, H. (1986a) Efficient statistical modelling of longitudinal data. *Annals of Human Biology*, **13**, 129–141.

*Goldstein, H. (1986b) Multilevel mixed linear model analysis using iterative generalized least-squares. *Biometrika*, **73**, 43–56 and 357–381.

*Goldstein, H. (1987) *Multilevel Models in Educational and Social Research*. Griffin, London.

Gould, A.L. (1980) A new approach to the analysis of clinical drug trials with withdrawals. *Biometrics*, **36**, 721–727.

Grassia, A. and De Boer, E.S. (1980) Some methods of growth curve fitting. *Mathematical Scientist*, **5**, 91–103.

*Greenhouse, S.W. and Geisser, S. (1959) On the methods in the analysis of profile data. *Psychometrika*, **24**, 95–112.

*Grieve, A.P. (1984) Tests of sphericity of normal distributions and the analysis of repeated measures designs. *Phychometrika*, **49**, 257–267.

*Grieve, A.P. (1985) A Bayesian analysis of the two-period crossover design for clinical trials. *Biometrics*, **41**, 979–990.

*Grizzle, J.E. (1965) The two period change-over design and its uses in clinical trials. *Biometrics*, **21**, 467–480. (1974) Corrigendum. *Biometrics*, **30**, 727.

Grizzle, J.E. (1970) An example of the analysis of a series of response curves and an application of multivariate multiple comparisons. Chapter 15 in *Essays in Probability and Statistics in the Memory of S.N. Roy* (eds R.C. Bose *et al.*). University of North Carolina Press, Chapel Hill, NC, pp. 311–326.

*Grizzle, J.E. and Allen, D.M. (1969) Analysis of growth and dose response curves. *Biometrics*, **25**, 357–381.

Guire, K.E. and Kowalski, C. (1979) Mathematical description and representation of developmental change functions in the intra- and inter-individuals levels. Chapter 4 in *Longitudinal Research in the Study of Behaviour and Development* (eds J.R. Nesselroade and P.B. Baltes). Academic Press, New York, pp. 89–110.

Guthrie, D. (1981) Analysis of dichotomous variables in repeated measures experiments. *Psychological Bulletin*, **90**, 189–195.

Hall, W.B. (1975) Repeated measurements experiments. In *Developments in Field Experimental Design Analysis* (eds V.J. Boxinger and J.L. Wheeler). Commonwealth Agricultural Bureau, pp. 33–34.

*Hand, D.J. and Taylor, C.C. (1987) *Multivariate Analysis of Variance and Repeated Measures*. Chapman and Hall, London.

*Harris, P. (1984) An alternative test for multisample sphericity. *Psychometrika*, **49**, 273–275.

*Hartley, H.O. (1967) Expectations, variances, and covariances of Anova mean squares by 'synthesis'. *Biometrics*, **23**, 105–114.

*Hartley, H.O. and Rao, J.N.K. (1967) Maximum likelihood estimation for the mixed model analysis of variance model. *Biometrika*, **54**, 93–108.

*Harville, D.A. (1975) Extension of the Gauss–Markov theorem to include the estimation of random effects. *Annals of Statistics*, **4**, 384–395.

Harville, D.A. (1976) Confidence intervals and sets for linear combintions of fixed and random effects. *Biometrics*, **32**, 403–407.

*Harville, D.A. (1977) Maximum likelihood approaches to variance component estimation and to related problems. *Journal of the American Statistical Association*, **72**, 320–340.

Healy, M.J.R. (1981) Some problems of repeated measurements, in *Perspectives in Medical Statistics* (eds Bithell and Coppie). Academic Press, New York.

Hearne, E.M., Clark, G.M. and Hatch, J.P. (1983) A test for serial correlation in univariate repeated-measure analysis. *Biometrics*, **39**, 237–243.

Heckman, J.J. (1981) The incidental parameters problem and the problem of initial conditions in estimating discrete time, discrete data stochastic processes. In *Structural Analysis of Discrete Data* (eds C.F. Manski and D. McFadden). MIT Press, Cambridge, MA.

Heckman, J.J. and Singer, B. (1984) A method for minimizing the impact of distributional assumptions in econometric models of duration. *Econometrica*, **52**, 271–320.

Heckman, J.J. and Willis, R.J. (1977) A beta-logistic model for the analysis of sequential labor force participation by married women. *Journal of Political Economy*, **85**, 27–58.

Hedayat, A. and Afsarinejad, K. (1975) Repeated measurements designs, I, in *A Survey of Statistical Design and Linear Models* (ed. J.N. Srivastava). North-Holland, Amsterdam, pp. 229–242.

Hedayat, A. and Afsarinejad, K. (1978) Repeated measures designs, II. *Annals of Statistics*, **6**, 619–628.

*Henderson, C.R., Kempthorne, O., Searle, S.R. and Von Krosigk, C.N. (1959) Estimation of environmental and genetic trends from records subject to culling. *Biometrics*, **15**, 192–218.

Hertzog, C. and Rovine, M. (1985) Repeated-measures analysis of variance in developmental research: selected issues. *Child Development*, **56**, 787–809.

*Hills, M. (1968) A note on the analysis of growth curves. *Biometrics*, **24**, 192–196.

*Hills, M. and Armitage P. (1979) The two-period cross-over clinical trial. *British Journal of Clinical Pharmacology*, **8**, 7–20.

Hoel, P.G. (1964) Methods for comparing growth type curves. *Biometrics*, **20**, 859–872.

Hudson, I.L. (1983) Asymptotic tests for growth curve models with autoregressive errors. *Australian Journal of Statistics*, **25**, 413–424.

Hui, S.L. (1984) Curve fitting for repeated measurements made at irregular time-points. *Biometrics*, **40**, 691–697.

*Huynh, H. (1978) Some approximate tests for repeated measurement designs. *Psychometrika*, **43**, 161–175.

Huynh, H. (1981) Testing the identity of trends under the restriction of monotonicity in repeated measures designs. *Psychometrika*, **46**, 295–305.

Huynh, H. and Feldt, L.S. (1970) Conditions under which mean square ratios in repeated measurements designs have exact F-distributions. *Journal of the American Statistical Association*, **65**, 1582–1589.

*Huynh, H. and Feldt, L.S. (1976) Estimation of the Box correction for degrees of freedom for sample data in randomised block and split-plot designs. *Journal of Educational Statistics*, **1**, 69–82.

Huynh, H. and Feldt, L.S. (1980) Performance of traditional *F* tests in repeated measures designs under covariance heterogeneity. *Communications in Statistics – Theory and Methods*, **A9**, 61–74.

*Huynh, H. and Mandeville, G.K. (1979) Validity conditions in repeated measures designs. *Psychological Bulletin*, **86**, 964–973.

Imhof, J.P. (1962) Testing the hypothesis of no fixed main-effects in Scheffé's mixed model. *Annals of Mathematical Statistics*, **33**, 1085–1095.

Imrey, P.B., Johnson, W.D. and Koch, G.G. (1975) An incomplete contingency table approach to paired comparison experiments. *Journal of the American Statistical Association*, **71**, 614–623.

Jennrich, R.I. (1977) Letting BMDP8V tell us something about randomised blocks repeated measures, and split-plots. *BMDP Statistical Software*, Technical Report No. 34, Department of Biomathematics, University of California, Los Angeles.

*Jennrich, R.I. and Schluchter, M.D. (1986) Unbalanced repeated-measures models with structured covariance matrices. *Biometrics*, **42**, 805–820.

Jensen, D.R. (1982) Efficiency and robustness in the use of repeated measurements. *Biometrics*, **38**, 813–825.

Johansen, S. (1982) Asymptotic inference in random coefficient regression models. *Scandinavian Journal of Statistics*, **9**, 201–207.

Johansen, S. (1984) Functional relations, random coefficients, and nonlinear regression with application to kinetic data. *Springer Lecture Notes in Statistics*, Vol. 22.

John, S. (1972) The distribution of a statistic used for testing sphericity of normal distributions. *Biometrika*, **59**, 169–173.

*Jones, B. and Kenward, M.G. (1987) Modelling binary data from a three period cross-over trial. *Statistics in Medicine*, **6**, 555–564.

*Jones, B. and Kenward, M.G. (1989) *Design and Analysis of Crossover Trials*. Chapman and Hall, London.

Jones, R.H. (1987) Serial correlation in unbalanced, mixed models. *Bulletin of the International Statistical Institute*, **46**.

Joreskog, K.G. (1979) Statistical estimation of structural models in longitudinal-developmental investigations. Chapter 11 in *Longitudinal Research in the Study of Behaviour and Development* (eds J.R. Nesselroade and P.B. Baltes). Academic Press, New York, pp. 303–351.

Jorgenson, M., Nielsen, C.T., Keiding, N. and Skakkebaek, N.E. (1985) *Parametric and non-parametric models applied to growth data*. Research Report, Statistical Research Unit, University of Copenhagen.

*Kalbfleisch, J.D. and Lawless, J.F. (1984) The estimation of transition probabilities from aggregate data. *Canadian Journal of Statistics*, **12**, 169–182.

*Kalbfleisch, J.D., Lawless, J.F. and Vollmer, W.M. (1983) Estimation in Markov models from aggregate data. *Biometrics*, **39**, 907–919.

Katz, B.M. and McSweeney, M. (1983) Some non-parametric tests for analysing ranked data in multi group repeated measures designs. *British Journal of Mathematical and Statistical Psychology*, **36**, 145–156.

Keen, A., Thissen, J.T.N.M., Hoeksta, J.A. and Jansen, J. (1986) Successive measurements experiments. *Statistica Neerlandica*, **40**, 205–223.

Kenny, D.A. and Judd, C.M. (1986) Consequences of violating the independence assumption in analysis of variance. *Psychological Bulletin*, **99**, 422–431.

*Kenward, M.G. (1979) An intuitive approach to the MANOVA test criteria. *The Statistician*, **28**, 193–198.

*Kenward, M.G. (1985) The use of fitted higher-order polynomial coefficients as covariates in the analysis of growth curves. *Biometrics*, **41**, 19–28.

Kenward, M.G. (1986) The distribution of a generalized least squares estimation with covariance adjustment. *Journal of Multivariate Analysis*, **20**, 244–250.

*Kenward, M.G. (1987) A method for comparing profiles of repeated measurements. *Applied Statistics*, **36**, 296–308.

*Kenward, M.G. and Jones B. (1987) A log linear model for binary and crossover data. *Applied Statistics*, **36**, 192–204.

Kenward, M.G. and Jones, B. (1988) Reply to Fidler 1988. *Applied Statistics*, **37**, 112–113.

*Kershner, R.P. and Federer, W.T. (1981) Two-treatment crossover designs for estimating a variety of effects. *Journal of the American Statistical Association*, **76**, 612–619.

*Keselman, H.J., Rogan, J.C., Mendoza, J.L. and Breen, L.J. (1980) Testing the validity conditions of repeated measures F tests. *Psychological Bulletin*, **87**, 479–481.

*Keselman, H.J. and Keselman, J.C. (1984) The analysis of repeated measures designs in medical research. *Statistics in Medicine*, **3**, 185–195.

Keuls, M. and Carretsen, F. (1982) Statistical analysis of growth curves in plant breeding. *Euphytica*, **31**, 51–64.

*Khatri, C.G. (1966) A note on a MANOVA model applied to problems in growth curves. *Annals of the Institute of Statistical Mathematics*, **18**, 75–86.

Khatri, C.G. and Shah, K.R. (1979) *Estimation in Non-linear Growth Curves*. Technical Report, University of Waterloo, Canada.

Kleinbaum, D.G. (1973) A generalisation of the growth curve model which allows missing data. *Journal of Multivariate Analysis*, **3**, 117–124.

Koch, G.G. (1969) Some aspects of the statistical analysis of 'split plot' experiments in completely randomized layouts. *Journal of the American Statistical Association*, **64**, 485–505.

Koch, G.G. (1970) The use of non-parametric methods in the statistical analysis of a complex split plot experiment. *Biometrics*, **26**, 105–128.

Koch, G.G. (1972) The use of nonparametric methods in the statistical analysis of the two-period change-over design. *Biometrics*, **28**, 577–584.

Koch, G.G., Amara, I.A., Stokes, M.E. and Gillings, D.B. (1980) Some views on parametric and non-parametric analyses for repeated measurements and selected bibliography. *International Statistical Review*, **48**, 249–265.

Koch, G.G., Gillings, D.B. and Stokes, M.E. (1980) Biostatistical implication of design, sampling and measurements to the analysis of health science data. *Annual Review of the Public Health*, **1**, 163–225.

Koch, G.G., Grizzle, J.E., Semenya, K. and Sen, P.K. (1978) Statistical methods for evaluation of mastitis treatment data. *Journal of Diary Science*, **61**, 829–847.

Koch, G.G., Imrey, P.B. and Reinfurt, D.W. (1972) Linear model analysis of categorical data with incomplete response vectors. *Biometrics*, **28**, 663–692.

*Koch, G.G., Landis, J.R., Freeman, J.L. Freeman, D.H. Jr. and Lehnen, R.G. (1977) A general methodology for the analysis of experiments with repeated measurement of categorical data. *Biometrics*, **33**, 133–158.

Koch, G.G. and Reinfurt, D.W. (1971) The analysis of categorical data from mixed models. *Biometrics*, **27**, 157–173.

Koch, G.G. and Sen, P.K. (1968) Some aspects of the statistical analysis of the mixed model. *Biometrics*, **24**, 27–47.

Kornegay, E.T., Mosanghini, V. and Snee, R.D. (1970) Urea and amino acid supplementation of swine diets. *Journal of Nutrition*, **100**, 330–340.

Kowalski, C.J. and Guire, K.E. (1974) Longitudinal data analysis. *Growth*, **38**, 131–169.

Koziol, J.A., Maxwell, D.A., Fukushima, M., Colmerauer, M.E.M. and Pilch, Y.H. (1981) A distribution-free test for a tumour-growth curve analysis with application to an animal tumour immunotherapy experiment. *Biometrics*, **37**, 383–390.

Krishnaiah, P.R. (1975) Simultaneous tests for multiple comparisons of growth curves when errors are autocorrelated. *Proceedings of the 40th Session of the International Statistical Institute*, **4**, 62–66.

Kunert, J. (1983) Optimal design and refinement of the linear model with applications to repeated measurements designs. *Annals of Statistics*, **1**, 247–257.

Kunert, J. (1984) Optimality of balanced uniform repeated measurements designs. *Annals of Statistics*, **12**, 1006–1017.

Kunert, J. (1985) Optimal repeated measurements designs for correlated observations and analysis by weighted least squares. *Biometrika*, **72**, 375–389.

Kunert, J. (1987) On variance estimation in crossover designs. *Biometrics*, **43**, 833–845.

La Vange, L.M. (1983) *Analysis of Incomplete Longitudinal Data with Constrained Covariance Structure*. Ph.D. dissertation, Department of Biostatistics, University of North Carolina, Chapel Hill.

*Laird, N.M., Lange, N. and Stram, D. (1987) Maximum likelihood computations with repeated measures: Application of the EM algorithm. *Journal of the American Statistical Association*, **82**, 97–105.

*Laird, N.M. and Ware, J.H. (1982) Random effects models for longitudinal data. *Biometrics*, **38**, 963–974.

Lancaster, A. and Nickell, S.J. (1980) The analysis of re-employment probabilities for the unemployed. *Journal of the Royal Statistical Society*, A, **143**, 141–165.

Landis, J.R. and Koch, G.G. (1977) The measurement of observer agreement for categorical data. *Biometrics*, **33**, 159–175.

Landis, J.R. and Koch, G.G. (1979) The analysis of categorical data in longitudinal studies of behavioural development. Chapter 9 in *Longitudinal Research in the Study of Behaviour and Development* (eds J.R. Nesselroade and P.B. Baltes). Academic Press, New York, pp. 233–261.

*Landis, J.R., Miller, M.E., Davis, C.S. and Koch, G.G. (1988) Some general methods for the analysis of categorical data in longitudinal studies. *Statistics in Medicine*, 7, 109–137.

Larsen, W.A. (1969) The analysis of variance for the two-way classification model with observations within a row serially correlated. *Biometrika*, 56, 509–515.

*Lawless, J.F. and McLeish, D.L. (1984) The information in aggregate data from Markov chains. *Biometrika*, 71, 419–430.

Layard, M.W.J. and Arvesen, J.N. (1978) Analysis of Poisson data in crossover experimental designs. *Biometrics*, 34, 421–428.

Lee, J.C. and Geisser, S. (1972) Growth curve prediction. *Sankhya* A, 34, 393–412.

Lee, J.C. and Geisser, S. (1975) Applications of growth curve prediction. *Sankhya* A, 37.

*Lee, T.C., Judge, G.G. and Zellner, A. (1977) *Estimating the Parameters of the Markov Probability Model from Aggregate Time Data.* North Holland, Amsterdam.

Lee, Y.H.K. (1974) A note on Rao's reduction of Potthoff and Roy's generalised linear model. *Biometrika*, 61, 349–351.

Leech, F.B. and Healy, M.J.R. (1959) The analysis of experiments on growth rate. *Biometrics*, 15, 98–106.

Leedow, M.I. and Tweedie, R.L. (1983) Weighted area techniques for the estimation of parameters of a growth curve. *Australian Journal of Statistics*, 25, 310–320.

Lehnen, R.G. and Koch, G.G. (1974) The analysis of experiments on growth rate. *Biometrics*, 15, 98–106.

Lehnen, R.G. and Koch, G.G. (1974) Analyzing panel data with uncontrolled attrition. *The Public Opinion Quarterly*, 38, 40–56.

Lehnen, R.G. and Koch, G.G. (1974) The analysis of categorical data from repeated measurement research designs. *Political Methodology*, 1, 103–123.

*Liang, K.Y. and Zeger, S.L. (1986) Longitudinal data analysis using generalised linear models. *Biometrika*, 73, 13–22.

Lienert, G.A. (1981) Defining and detecting second order correlations in repeated measurement designs. *Biometrical Journal*, **23**, 167–173.

Lienert, G.A. and Netter, P. (1986) Nonparametric evaluation of repeated measurement designs by points-symmetry testing. *Biometrical Journal*, **28**, 3–10.

Liski, E.P. (1983) Curve fitting for repeated measurements – an application to longitudinal data. *Acta Universitas Tamerensis*, A, **153**, 62–75.

Liski, E.P. (1987) A growth curve analysis for bulls tested at station. *Biometrical Journal*, **29**, 331–343.

Longford, N. (1985) Intrasubject covariance estimation in longitudinal studies. *Biometrics*, **41**, 1075–1076.

*Longford, N.T. (1987) A fast scoring algorithm for maximum likelihood estimation in unbalanced mixed models with nested random effects. *Biometrika*, **74**, 817–827.

Lucas, H.L. (1951) Bias estimation of error in change-over trials with dairy cattle. *Journal of Agricultural Science*, **41**, 146–148.

Lyung, G.M. and Box, G.E.P. (1980) Analysis of variance with autocorrelated observations. *Scandinavian Journal of Statistics*, **7**, 172–180.

Machin, D. (1975) On a design problem in growth studies. *Biometrics*, **31**, 749–753.

*Madansky, A. (1959) Least squares estimation in finite Markov processes. *Psychometrika*, **24**, 137–144.

Madsen, K.W. (1977) A growth curve model for studies in morphometrics. *Biometrics*, **33**, 659–669.

Madssen, K.S., Meller, J.P. and Province, M.A. (1986) The use of an extended baseline period in the evaluation of treatment in a longitudinal Duchenne muscular dystrophy trial. *Statistics in Medicine*, **5**, 231–241.

Mager, P.P. (1973) Repeated measurement designs. *Activitas Nervosa Superior*, **15**, 269–277.

Mak, T.K. and Ng, K.W. (1981) Analysis of familial data: linear model approach. *Biometrika*, **68**, 457–461.

Mansour, H., Nordheim, E.V. and Rutledge, J.J. (1985) Maximum likelihood estimation of variance components in repeated measures designs assuming autoregressive errors. *Biometrics*, **41**, 287–294.

Marks, E. (1968) Some profile methods useful with singular covariance matrices. *Psychological Bulletin*, **70**, 179–184.

Massy, W.F., Montgomery, D.B., and Morrison, D.G. (1970) *Stochastic Models of Buying Behaviour.* MIT Press, Cambridge, MA.

*Matthews, J.N.S. (1987) Optimal crossover designs for the comparison of two treatments in the presence of carryover effects and autocorrelated errors. *Biometrika,* **74**, 311–320.

*Mauchly, J.W. (1940) Significance test for sphericity of a normal *n*-variate distribution. *Annals of Mathematical Statistics,* **29**, 204–209.

McCall, R.B. and Appelbaum, M.I. (1973) Bias in the analysis of repeated-measures designs: some alternative approaches. *Child Development,* **44**, 401–415.

McCammon, R.W. (1970) *Human Growth and Development.* Thomas, Springfield.

McLeish, D.L. (1984) Estimation for aggregate models: The aggregate Markov chain. *Canadian Journal of Statistics,* **12**, 265–282.

*McMahan, C.A. (1981) An index of tracking. *Biometrics,* **37**, 447–455.

*Mead, R. (1970) The non-orthogonal design of experiments. *Applied Statistics,* **19**, 64–81.

Mendoza, J.L. (1980) A significance test for multisample sphericity. *Psychometrika,* **45**, 495–498.

*Mendoza, J.L., Toothaker, L.E., and Nicewander, W.A. (1974) A Monte Carlo comparison of the univariate and multivariate methods for the groups by trials repeated measures design. *Multivariate Behavioural Research,* **9**, 165–178.

Mendoza, J.L., Toothaker, L.E., and Crain, B.R. (1976) Necessary and sufficient conditions for F ratios in the $L \times J \times K$ factorial design with two repeated factors. *Journal of the American Statistical Association,* **71**, 992–993.

*Miller, G.A. (1952) Finite Markov processes in psychology. *Psychometrika,* **17**, 149–167.

*Miller, J.J. (1977) Asymptotic properties of maximum likelihood estimators in the mixed model analysis of variance. *Annals of Statistics,* **5**, 746–762.

*Mitzel, H.C. and Games, P.A. (1981) Circularity and multiple comparisons in repeated measures designs. *British Journal of Mathematical and Statistical Psychology,* **34**, 253–259.

*Monlezun, C.J., Blouin, D.C. and Malone, L.C. (1984)

Contrasting split plot and repeated measures experiments and analyses. *The American Statistician*, **38**, 21–31.

Morrison, D.F. (1970) The optimal spacing of repeated measurements. *Biometrics*, **26**, 281–290.

Morrison, D.F. (1972) The analysis of a single sample of repeated measurements. *Biometrics*, **28**, 55–71.

*Moulton, L.H. and Zeger, S.L. (1989) Analyzing repeated measures on generalized linear models via the bootstrap. *Biometrics*, **45**, 381–394.

*Nagarsenker, B.N. and Pillai, K.C.S. (1972) *The distribution of the sphericity test criterion (ARL 72-0154)*. Wright Patterson Air Force Base, Ohio: Aerospace Research Laboratory, November. (NTIS number AD-754 232.)

Nagarsenker, B.N. and Pillai, K.C.S. (1973) The distribution of the sphericity test criterion. *Journal of Multivariate Analysis*, **3**, 226–235.

Nicholls, A.O. and Calder, D.M. (1973) Comments on the use of regression analysis for the study of plant growth. *New Phytology*, **72**, 571–581.

O'Brien, R.G. and Kaiser, M.K. (1985) MANOVA method for analyzing repeated measures designs: an extensive primer. *Psychological Bulletin*, **97**, 316–333.

Ochi, Y. and Prentice, R.L. (1984) Likelihood inference in a correlated probit regression model. *Biometrika*, **71** 531–543.

Olkin, I. and Vaeth, M. (1981) Maximum likelihood estimation in a two-way analysis of variance with correlated errors in one classification. *Biometrika*, **68**, 653–660.

Palta, M. and Cook, T. (1986) *Some considerations in the analysis of rates of change in longitudinal studies*. Research report, Wisconsin Clinical Cancer Center, Biostatistics, University of Wisconsin, Madison.

Pantula, S.G. and Pollock, K.H. (1985) Nested analysis of variance with autocorrelated errors. *Biometrics*, **41**, 909–920.

Patel, H.I. (1983) Use of baseline measurements in two-period crossover design in clinical trials. *Communications in Statistics – Theory and Methods*, **12**, 2693–2712.

Patel, H.I. and Hearne, E.M. (1980) Multivariate analysis for the two-period repeated measures crossover design with application to clinical trials. *Communications in Statistics – Theory and Methods*, **9**, 1919–1929.

*Patel, H.I. and Khatri, C.G. (1981) Analysis of incomplete data in experiments with repeated measurements in clinical trials using a stochastic model. *Communications in Statistics – Theory and Method*, **10**, 2259–2277.

Patterson, H.D. (1950) The analysis of change-over trials. *Journal of Agricultural Science*, **40**, 375–380.

Patterson, H.D. (1951) Change-over trials. *Journal of the Royal Statistical Society*, B, **13**, 256–271.

Patterson, H.D. and Lowe, B.I. (1970) The errors of long-term experiments. *Journal of Agricultural Science, Cambridge*, **74**, 53–60.

Patterson, H.D. and Thompson, R. (1974) Maximum likelihood estimation of components of variance. *Proceedings of the 8th International Biometric Conference*, 197–207.

Payne, R.W. and Richardson, M.G. (1981) Repeated measures in GENSTAT. *ASA Pro. St. Cp, 30–38.*

Pettitt, A.N. (1984) Fitting a sinusoid to biological rhythm data using ranks. *Biometrics*, **40**, 295–300.

Plewis, I. (1981) A comparison of approaches to the analysis of longitudinal categorical data. *British Journal of Mathematical and Statistical Psychology*, **34**, 118–123.

Plewis, I. (1985) *Analysing Change: measurement and explanation using longitudinal data.* Wiley, New York.

Poor, D.D.S. (1973) Analysis of variance for repeated measures designs: two approaches. *Psychological Bulletin*, **80**, 204–209.

*Potthoff, R.R. and Roy S.N. (1964) A generalized multivariate analysis of variance model useful especially for growth curve problems. *Biometrika*, **51**, 313–326.

*Prescott, R.J. (1981) The comparison of success rates in cross-over trials in the presence of an order effect. *Applied Statistics*, **30**, 9–15.

Pruitt, Y.M., De Muth, R.E. and Turner, M.E. (1979) Practical application of genetic growth theory and the significance of growth curve parameters. *Growth*, **43**, 19–35.

*Racine-Poon, A. (1985) A Bayesian approach to nonlinear random effects models. *Biometrics*, **41**, 1015–1023.

Rao, C.R. (1958) Some statistical methods for comparison of growth curves. *Biometrics*, **14**, 1–17.

*Rao, C.R. (1965) The theory of least-squares when the parameters are stochastic and its application to the analysis of growth curves. *Biometrika*, **52**, 447–458.

*Rao, C.R. (1975) Simultaneous estimation of parameters in different linear models and applications to biometric problems. *Biometrics*, **31**, 545–555.

*Rao, J.N.K. (1968) On expectations, variances and covariances of Anova mean squares by 'synthesis'. *Biometrics*, **24**, 963–978.

Rao, M.N. and Rao, C.R. (1966) Linked cross sectional study for determining norms and growth rates – a pilot survey of Indian school going boys. *Sankhya* B, **28**, 237–258.

*Reinsel, G. (1982) Multivariate repeated-measurement or growth curve models with multivariate random-effects covariance structure. *Journal of the American Statistical Association*, **77**, 190–195.

Reinsel, G. (1984) Estimation and prediction in a multivariate random effects generalized linear model. *Journal of the American Statistical Association*, **79**, 406–414.

*Reinsel, G. (1985) Mean squared error properties of empirical Bayes estimators in a multivariate random effects general linear model. *Journal of the American Statistical Association*, **80**, 642–650.

Richards, F.J. (1959) A flexible growth function for empirical use. *Journal of Experimental Biology*, **10**, 190–200.

Roberts, E.A. (1980) Stratification, covariance and weight gain in experiments with repeated observations. *Journal of Agricultural Science, Cambridge*, **95**, 141–145.

Roberts, E.A. and Raison, J.M. (1983) An analysis of moisture content of soil cores in a designed experiment. *Biometrics*, **39**, 1097–1105.

*Rochon, J. and Helms, R.W. (1989) Maximum likelihood estimation for incomplete repeated measures experiments under an ARMA covariance structure. *Biometrics*, **45**, 207–218.

*Rogan, J.C., Keselman, H.J. and Mendoza, J.L. (1979) Analysis of repeated measurements. *British Journal of Mathematical and Statistical Psychology*, **32**, 269–286.

*Rosenberg, B. (1973) Linear regression with randomly dispersed parameters. *Biometrika*, **60**, 65–72.

*Rouanet, H. and Lepine, D. (1970) Comparison between treatments in a repeated-measures design: ANOVA and multivariate methods. *British Journal of Mathematical and Statistical Psychology*, **23**, 147–163.

*Rowell, J.G. and Walters, R.E. (1976) Analysing data with repeated observations on each experimental unit. *Journal of Agricultural Science, Cambridge*, **87**, 423–432.

Sallas, W.M. and Harville, D.A. (1981) Best linear recursive estimation for mixed linear models. *Journal of the American Statistical Association*, **76**, 860–869.

Salsburg, D. (1981) Development of statistical analysis for single dose bronchodilators. *Controlled Clinical Trials*, **2**, 305–317.

Sandland, R.L. (1983) Mathematics and growth of organisms. *Mathematical Scientist*, **5**, 11–30.

*Sandland, R.L. and McGilchrist, C.A. (1979) Stochastic growth curve analysis. *Biometrics*, **35**, 255–272.

*Schaeffer, L.R., Wilton, J.W. and Thompson, R. (1978) Simultaneous estimation of variance and covariance components from multitrait mixed model equations. *Biometrics*, **34**, 199–208.

Scheffé, H. (1956a) A mixed model for the analysis of variance. *Annals of Mathematical Statistics*, **27**, 23–26.

Scheffé, H. (1956b) Alternative models for the analysis of variance. *Annals of Mathematical Statistics*, **27**, 251–271.

Schill, W. and Louton (1980) On the analysis of variance for time sequenced samples. *Biometrical Journal*, **22**, 675–682.

Schwertman, N.C. (1978) A note on the Geisser–Greenhouse correction for incomplete data split-plot analysis. *Journal of the American Statistical Association*, **73**, 393–396.

Schwertman, N.C., Fridshal, D.A. and Magrey, J.M. (1981) On the analysis of incomplete growth curve data, a Monte Carlo study of the two nonparametric procedures. *Communications in Statistics – Simulation and Computation*, B, **10**, 51–66.

Searle, S.R. (1974) Prediction, mixed models, and variance components, in *Reliability and Biometry* (eds F. Proschan and R.J. Serfling). Society for Industrial and Applied Mathematics, Philadelphia, pp. 229–266.

Shaffer, J.P. (1981) The analysis of variance mixed model with allocated observations: application to repeated measurement designs. *Journal of the American Statistical Association*, **76**, 607–611.

Snee, R.D. (1972a) On the analysis of response curve data. *Technometrics*, **14**, 47–61.

Snee, R.D. (1972b) A useful method for conducting carrot shape studies. *Journal of Horticulture Science*, **47**, 267–277.

Snee, R.D. and Andrews, H.P. (1971) Statistical design and analysis of shaped studies. *Journal of the Royal Statistical Society*, C, **20**, 250–258.

Snee, R.D., Acuff, S.K. and Gibson, J.R. (1979) A useful method for the analysis of growth studies. *Biometrics*, **35**, 835–848.

Spjotvoll, E. (1977) Random coefficient regression models. A review. *Math. Operationsforsch Statist. Ser. Statistics*, **8**, 69–93.

Sprent, P. (1965) Fitting a polynomial to correlated equally spaced observations. *Biometrika*, **52**, 75–76.

Sprent, P. (1967) Estimation of mean growth curves for groups of organisms. *Journal of Theoretical Biology*, **17**, 159–173.

*Srivastava, M.S. (1984) Estimation of interclass correlations in familial data. *Biometrika*, **71**, 177–185.

*Srivastava, M.S. and Keen, K.J. (1988) Estimation of the interclass correlation coefficient. *Biometrika*, **75**, 731–739.

*Srivastava, M.S. and McDonald, L.L. (1974) Analysis of growth curves under the hierarchical models. *Sankhya* A, **36**, 251–260.

Stanish, W.M., Gillings, D.B. and Koch, G.G. (1978) An application of multivariate ratio methods to a longitudinal clinical trial with missing data. *Biometrics*, **34**, 305–317.

*Stiratelli, R., Laird, N. and Ware, J.H. (1984) Random effects models for serial observations with binary response. *Biometrics*, **40**, 961–971.

*Stoloff, P.H. (1970) Correcting for heterogeneity of covariance for repeated measures designs of the analysis of variance. *Educational and Psychological Measurement*, **30**, 909–924.

Stram, D.O., Wei, L.J. and Ware, J.H. (1988) Analysis of repeated ordered categorical outcomes with possible missing observations and time-dependent covariates. *Journal of the American Statistical Association*, **83**, 631–637.

Strenio, J.G., Weisberg, H.I. and Bryk, A. (1983) Empirical Bayes estimation of individual growth-curve parameters and their relationship to covariates. *Biometrics*, **39**, 71–86.

*Swamy, P.A.V.B. (1970) Efficient inference in a random coefficient regression model. *Econometrika*, **38**, 311–323.

Swamy, P.A.V.B. (1971) Statistical inference in random coefficient regression models. *Lecture Notes in Operational Research and Mathematical Systems*, Vol. 55. Springer-Verlag, New York.

Sweeting, T.J. (1982) A Bayesian analysis of some pharmacological data using a random coefficient regression model. *Applied Statistics*, **31**, 205–213.

Sweeting, T.J. (1984) The Bayesian analysis of random coefficient regression models. *Journal of the National Academy of Mathematics*, **2**, 67–80.

Taka, W.T. and Armitage, P. (1983) Autoregressive models in clinical trials. *Communications in Statistics – Theory and Methods*, **12**, 865–876.

*Thomas, D.R. (1983) Univariate repeated measures techniques applied to multivariate data. *Psychometrika*, **48**, 451–464.

Timm, M.H. (1980) Multivariate analysis of variance of repeated measures. In *Handbook of Statistics*, Vol: 1 (ed. P.R. Krishnaiah). North Holland, Amsterdam, pp. 42–47.

Tubbs, J.D., Lewis, T.O. and Duran, B.S. (1975) A note on the analysis of the MANOVA model and its application to growth curves. *Communications in Statistics – Theory and Methods*, A, **4**, 643–653.

*Van der Plas (1983) On the estimation of the parameters of Markov probability models using macro data. *Annals of Statistics*, **11**, 78–85.

Vanhonacker, W.R. (1984) Finite sample efficiency in a general linear model with serially correlated disturbances: A comparison of estimators given both temporally aggregated and disaggregated data. *Journal of Statistical Computation and Simulation*, **19**, 185–203.

Verbyla, A.P. (1986) Conditioning in the growth curve model. *Biometrika*, **73**, 475–483.

*Verbyla, A.P. and Venables, W.N. (1988) An extension of the growth curve model. *Biometrika*, **75**, 129–138.

Vonesh, E.F. and Schork, M.A. (1986) Sample sizes in the multivariate analysis of repeated measurements. *Biometrics*, **42**, 601–610.

Von Rosen, D. (1985) Multivariate linear normal models, with special reference to the growth curve model. *Doktordisputats*, Stockholms Universitat.

Wallenstein, S. (1982) Regression models for repeated measurements. *Biometrics*, **38**, 849–853.

*Wallenstein, S. and Fisher, A.C. (1977) The analysis of the two-period repeated measurements crossover design with applications to clinical trials. *Biometrics*, **33**, 261–269.

Wallenstein, S. and Fleiss, J.L. (1979) Repeated measurements analysis of variance when the correlations have a certain pattern. *Psychometrika*, **44**, 229–233.

*Walters, D.E. and Rowell, J.G. (1982) Comments on a paper by I. Olkin and M. Vaeth on two-way analysis of variance with correlated errors. *Biometrika*, **69**, 664–666.

*Ware, J.H. (1985) Linear models for the analysis of longitudinal studies. *American Statistician*, **39**, 95–101.

Ware, J.H., Lipsitz, S. and Speiger, F.E. (1988) Issues in the analysis of repeated categorical outcomes. *Statistics in Medicine*, **7**, 95–107.

*Ware, J.H. and Wu, M.C. (1981) Tracking: prediction of future values from serial measurements. *Biometrics*, **37**, 427–437.

Watson, G.S. (1955) Serial correlation in regression analysis. I. *Biometrika*, **42**, 327–341.

*Wei, L.J. and Johnson, W.E. (1985) Combining dependent tests with incomplete repeated measurements. *Biometrika*, **72**, 359–364.

Wei, L.J. and Stram, D. (1988) Analyzing repeated measurements with possibly missing observations by modelling marginal distributions. *Statistics in Medicine*, **7**, 139–148.

Wilk, M.B. and Kempthorne, O. (1955) Fixed, mixed and random models in the analysis of variance. *Journal of the American Statistical Association*, **50**, 1144–1167.

*Willan, A.R. and Pater, J.L. (1986) Carryover and the two period crossover clinical trial. *Biometrics*, **41**, 979–990.

Wilson, A.L. and Douglas, A.W. (1969) A note on non linear growth curve fitting. *American Statistician*, **23**, 37–38.

Wilson, B.J. (1977) Growth curves, their analysis and use, in *Growth and Poultry Meat Production* (eds K.N. Boorman and B.J. Wilson). British Poultry Science Ltd, Edinburgh, pp. 89–115.

*Wilson, K. (1975) The sampling distributions of conventional, conservative and corrected F-ratios in repeated measurements designs with heterogeneity of covariance. *Journal of Statistical Computing and Simulation*, **3**, 201–215.

*Wilson, P.D., Hebel, J.R. and Sherwin, R. (1981) Screening and diagnosis when within-individual observations are Markov-dependent. *Biometrics*, **37**, 553–565.

*Wishart, J. (1938) Growth rate determination in nutrition studies with the bacon pig, and their analysis. *Biometrika*, **30**, 16–28.

Woolson, R.F., Leeper, J.D. and Clarke, W.R. (1978) Analysis of incomplete data from longitudinal and mixed longitudinal studies. *Journal of the Royal Statistical Society*, A, **141**, 242–252.

Woolson, R.F. and Leeper, J.D. (1980) Growth curve analysis of complete and incomplete longitudinal data. *Communications in Statistics*, A, 1491–1513.

Woolson, R.F. and Clarke, W.R. (1984) Analysis of categorical incomplete longitudinal data. *Journal of the Royal Statistical Society*, A, **147**, 87–99.

*Yates, F. (1982) Regression models for repeated measurements. *Biometrics*, **38**, 850–853.

Young, C.K.M. and Carter, R.L. (1983) One-way analysis of variance with time-series data. *Biometrics*, **39**, 747–751.

Zeger, S.L. (1988a) The analysis of discrete longitudinal data: commentary. *Statistics in Medicine*, 7, 161–168.

*Zeger, S.L. (1988b) A regression model for time series of counts. *Biometrika*, **75**, 621–629.

*Zeger, S.L. and Liang, K.Y. (1986) Longitudinal data analysis for discrete and continuous outcomes. *Biometrics*, **42**, 121–130.

*Zeger, S.L. and Qaqish, B. (1988) Markov regression models for time series: a quasi-likelihood approach. *Biometrics*, **44**, 1019–1031.

*Zeger, S.L., Liang, K.Y. and Albert, P.S. (1988) Models for longitudinal data: a generalized estimating equation approach *Biometrics*, **44**, 1049–1060.

Zeger, S.L., Liang, K.Y. and Steven, G.S. (1985) The analysis of binary longitudinal data with time-independent covariates. *Biometrika*, **72**, 31–38.

Zellner, A. and Tiao, G.C. (1964) Bayesian analysis of the regression model with autocorrelated errors. *Journal of the American Statistical Association*, **59**, 763–778.

Zerbe, G.O. and Walker, S.H. (1977) A randomization test for comparison of groups of growth curves with different polynomial design matrices. *Biometrics*, **33**, 653–657.

Zerbe, G.O. (1979) Randomization analysis of the completely randomised design extended to growth and response curves. *Journal of the American Statistical Association*, **74**, 215–221.

Zerbe, G.O. and Jones, R.H. (1980) An application of growth curve techniques to time series data. *Journal of the American Statistical Association* **75**, 507–509.

Zimmerman, H. and Rahlfs V. (1978) Testing hypotheses in the two-period change-over with binary data. *Biometrical Journal*, **20**, 133–141.

Supplementary references

Andersen, E.B. (1973) *Conditional Inference and Models for Measuring.* Mentalhygiejnisk Forlag, Copenhagen.

Andersen, E.B. (1980) *Discrete Statistical Models with Social Science Applications.* Amsterdam, New York.

Anderson, R.L. and Bancroft, T.A. (1952) *Statistical Theory in Research.* McGraw-Hill, New York.

Anderson, T.W. (1969) Statistical inference for covariance matrices with linear structure, in *Multivariate Analysis II* (ed. P. Krishnaiah). Academic Press, New York.

Anderson, T.W. (1970) Estimation of covariance matrices which are linear combinations or whose inverses are linear combinations of given matrices, in *Essays in Probability and Statistics* (eds R.C. Bose et al.). University of North Carolina Press, Chapel Hill, pp. 1–24.

Anderson, T.W. (1973) Asymptotically efficient estimation of covariance matrices with linear structure. *Analysis of Statistics*, **1**, 135–141.

Anderson, T.W. (1984) *An Introduction to Multivariate Statistical Analysis.* Wiley, New York.

Andrade, D.F. and Helms, R.W. (1984) Maximum likelihood estimates in the multivariate normal with patterned mean and covariance via the EM algorithm. *Communications in Statistics – Theory and Methods*, **13**, 2239–2251.

Armitage, P. (1982) Reflections on some trends in medical statistics. *Utilitas Mathematica*, A, **21**, 101–117.

Arnold, S.F. (1981) *The Theory of Linear Models and Multivariate Analysis.* Wiley, New York.

Atkinson, A.C. (1985) *Plots, Transformations and Regression.* Clarendon Press, Oxford.

Barnett, V. and Lewis, T. (1978) *Outliers in Statistical Data.* Wiley, New York.

Barlett, M.S. (1937) The statistical conception of mental factors. *British Journal of Psychology*, **28**, 97–104.

Bartlett, M.S. (1937) Some examples of statistical methods of research in agriculture. *Journal of the Royal Statistical Society Supplement*, **4**, 137–183.

Battese, G.E. and Fuller, W.A. (1972) Determination of economic optima from crop rotation experiments. *Biometrics*, **28**, 781–792.

Beale, E.M.L. and Little, R.J.A. (1975) Missing values in multivariate analysis. *Journal of the Royal Statistical Society*, **B**, **37**, 129–145.

Beauchamp, J.J. and Hoel, D.G. (1974) Some investigation and simulation studies of economical analysis. *Journal of Statistical Computation and Simulation*, **2**, 197–209.

Berenblut, I.I. and Webb, G.I. (1973) A new test for autocorrelated errors in the linear regression model. *Journal of the Royal Statistical Society*, B, **35**, 33–50.

Bhapkar, V.P. and Somes, G.W. (1977) Distribution of Q when testing equality of matched proportions. *Journal of the American Statistical Association*, **72**, 658–661.

Bishop, Y.M.M., Fienberg, S.E. and Holland, P.W. (1975) *Discrete Multivariate Analysis*. MIT Press, Cambridge, MA.

Bock, R.D. (1975) *Multivariate Statistical Methods in Behavioural Research*. McGraw-Hill, New York.

Bock, R.D. and Aitken, M. (1981) Marginal maximum likelihood estimation of item parameters: application of the EM algorithm. *Psychometrika*, **46**, 443–59.

Box, G.E.P. (1949) A general distribution theory for a class of likelihood criteria. *Biometrika*, **36**, 317–346.

Box, G.E.P. (1954a) Some theorems on quadratic forms applied in the study of analysis of variance problems: I, Effects of inequality of variance in the one-way classification. *Annals of Mathematical Statistics*, **25**, 290–302.

Box, G.E.P. (1954b) Some theorems on quadratic forms applied in the study of analysis of variance problems: II, Effects of inequality of variance in the two-way classification. *Annals of Mathematical Statistics*, **25**, 484–497.

Box, G.E.P. and Cox, D.R. (1964) An analysis of transformations (with discussion). *Journal of the Royal Statistical Society*, B, **26**, 211–252.

Box, M.J. (1971) Bias in nonlinear estimation (with discussion). *Journal of the Royal Statistical Society*, B, **33**, 171–201.

Brewer, A.C. and Mead, R. (1986) Continuous second order models of spatial variation with application to the efficiency of field crop experiments (with discussion). *Journal of the Royal Statistical Society*, A, **149**, 314–348.

Brown, M.B. (1975) A method for combining non-independent, one-sided tests of significance. *Biometrics*, **31**, 987–992.

Chaloner, K. and Brant, R. (1988) A Bayesian approach to outlier detection and residual analysis. *Biometrika*, **75**, 651–660.

Chatterjee, S.K. and Sen, P.K. (1966) Nonparametric tests for the multisample multivariate location problem, in *Essays in Probability and Statistics* (eds R.C. Bose, I.M. Chakravorti, B.C. Mahalanobis, C.R. Rao, and K.J.C. Smith). University of North Carolina Press, pp. 197–228.

Cochran, W. (1950) The comparison of percentages in matched samples. *Biometrika*, **37**, 256–266.

Cochran, W.G. and Bliss, C.I. (1948) Discriminant functions with covariance. *Annals of Mathematical Statistics*, **19**, 151–176.

Cochran, W.G. and Cox, G. (1957) *Experimental Designs*, 2nd edn. Wiley, New York.

Conover, W.J. (1971) *Practical Nonparametric Statistics*. Wiley, New York.

Cook, R.D. and Weisberg, S. (1982) *Residuals and Influence in Regression*. Chapman and Hall, London.

Cox, D.R. and Oakes, D. (1984) *Analysis of Survival Data*. Chapman and Hall, London.

Cox, D.R. and Snell, E.J. (1968) A general definition of residuals (with discussion). *Journal of the Royal Statistical Society*, B, **30**, 248–275.

Cox, M.A.A. and Plackett, R.L. (1980) Matched pairs in factorial experiments with binary data. *Biometrical Journal*, **22**, 697–702.

Crowder, M.J. (1985) Gaussian estimation for correlated binomial data *Journal of the Royal Statistical Society*, B, **47**, 229–237.

Crowder, M.J. (1987) On linear and quadratic estimating functions. *Biometrika*, **74**, 591–597.

Daniels, H.E. (1938) The effect of departures from ideal conditions other than non-normality on the t and Z tests of significance. *Proceedings of the Cambridge Philosophical Society*, **34**, 321–328.

Dawid, A.P. (1988) Symmetry models and hypotheses for structured data layouts (with discussion). *Journal of the Royal Statistical Society*, B, **50**, 1–34.

Dempster, A.P. (1972) Covariance selection. *Biometrics*, **28**, 157–175.

Dempster, A.P., Laird, N.M. and Rubin, D.B. (1977) Maximum likelihood from incomplete data via the EM algorithm (with discussion). *Journal of the Royal Statistical Society*, B, **39**, 1–38.

Dixon, W.J. (ed.) (1983) *BMDP Statistical Software*. University of California Press, Berkeley.

Draper N.R. and Smith H. (1966) *Applied Regression Analysis*. Wiley, New York.

Eisenhart, C. (1947) The assumptions underlying the analysis of variance. *Biometrics*, **3**, 1–21.

Feder, P.I. (1975) On the asymptotic distribution theory in segmented regression problems-identified case. *Annals of Statistics*, **3**, 49–83.

Federer, W.T. (1955) *Experimental Design*. The MacMillan Company, New York.

Finn, J.D. and Mattson, I. (1978) *Multivariate Analysis in Educational Research: applications of the MULTIVARIANCE program*. National Educational Resources, Chicago.

Fleiss, J.L. (1973) *Statistical Methods for Rates and Proportions*. Wiley, New York.

Fletcher, R. and Powell, M.J.D. (1963) A rapidly convergent descent method for minimization. *Computer Journal*, **6**, 163–168.

Friedman, M. (1937) The use of ranks to avoid the assumption of normality implicit in the analysis of variance. *Journal of the American Statistical Association*, **32**, 675–699.

Gafarian, A.V. (1978) Confidence bands in multivariate polynomial regression. *Technometrics*, **20**, 141–149.

Geisser, S. (1971) The inferential use of predictive distributions, in *Foundations of Statistical Inference*. Holt, Rinehart and Winston, pp. 456–469.

Gerig, T. (1969) A multivariate extension of Friedman's test. *Journal of the American Statistical Association*, **64**, 1595–1608.

Gibaldi, M. and Perrier, D. (1975). *Pharmacokinetics*. Marcel Dekker, New York.

Gill, J.L. (1978) *Design and Analysis of Experiments in the Animal and Medical Sciences*. Iowa State University Press, Ames, Iowa.

Glasser, M. (1965) Regression analysis with dependent variable censored. *Biometrics*, **21**, 300–307.

Gleser, L.J. and Olkin, I. (1966) A 'k'-sample regression model with covariance, in *Multivariate Analysis* (ed. P.R. Krishnaiah). Academic Press, New York, pp. 59–72.

Gleser, L.J. and Olkin, I. (1970) Linear models in multivariate analysis, in *Essays in Probability and Statistics* (eds Bose *et al.*) University of North Carolina Press, Chapel Hill, pp. 267–292.

Gleser, L.J. and Olkin, I. (1972) Estimation for a regression model with an unknown covariance matrix. *Proceedings of the Sixth*

Berkeley Symposium on Mathematical Statistics and Probability, **2**, 541–568.

Gokhale, D.V. and Kulback, S. (1978) *The Information in Contingency Tables*. Marcel Dekker, New York.

Greenblatt, D.J. and Koch-Weser, J. (1975) Clinical pharmacokinetics. *New England Journal of Medicine*, **293**, 702–704.

Grizzle, J.E., Williams, O.D. and Woolson, R.F. (1975) Multivariate analysis in biometry, in *Proceedings of the 8th International Biometrics Conference* (eds L.C.A. Corsten and T. Postelnicu). Editure Academiei Republicii Socialiste Romania, pp. 73–84.

Groggel, D.J. and Skillings, J.H. (1986) Distribution-free tests for main effects in multifactor designs. *The American Statistician*, **40**, 99–102.

Halperin, M. (1963) Confidence interval estimation in nonlinear regression. *Journal of the Royal Statistical Society*, B, **25**, 330–333.

Harris, R.J. (1975) *A Primer of Multivariate Statistics*. Academic Press, New York.

Hartree, D.R. (1958) *Numerical Analysis*, 2nd edn. Oxford University Press.

Heckman, J.J. (1979) Sample selection bias as a specification error. *Econometrica*, **47**, 143–162.

Henk Don, F.J. and Magnus, J.R. (1980) On the unbiasedness of iterated GLS estimators. *Communications in Statistics – Theory and Methods*, **A9**, 519–527.

Hettmansperger, T.P. and McKean, J.W. (1973) On testing for significant change in $c \times c$ tables. *Communications in Statistics – Theory and Methods*, **2**, 551–560.

Hinde, J. (1982) Compound Poisson regression models, in *GLIM82* (ed. R. Gilchrist). Springer-Verlag, New York, pp. 109–121.

Hinkley, D.V. (1988) Bootstrap methods. *Journal of the Royal Statistical Society*, B, **151**, 321–337.

Hochberg, Y. (1988) A sharper Bonferroni procedure for multiple tests of significance. *Biometrika*, **75**, 800–802.

Hochberg, Y. and Tamhane, A. (1987) *Multiple Comparison Procedures*. Wiley, New York.

Hollander, M. and Wolfe, D.A. (1973) *Nonparametric Statistical Methods*. Wiley, New York.

Hopkins, J.W. and Clay, P.P.F. (1963) Some empirical distributions of bivariate T^2 and homoscedasticity criterion M under unequal variance and leptokurtosis. *Journal of the American Statistical Association*, **58**, 1048–1053.

Horton, R.L. (1978) *The General Linear Model*. McGraw-Hill, New York.

Hotelling, H. (1947) Multivariate quality control, illustrated by the air testing of sample bombsights, in *Techniques of Statistical Analysis* (eds C. Eisenhart, M. Hastay and W.A. Wallis). McGraw-Hill, New York, pp. 111–184.

Hotelling, H. (1951) A generalized *T* test and measure of multivariate dispersion, in *Proceedings of the Second Berkeley Symposium on Mathematical Statistics and Probability* (ed. J. Neyman). University of California Press, pp. 23–41.

Ireland, C.T., Ku, H.H. and Kullback, S. (1969) Symmetry and marginal homogeneity of an *r* × *c* contingency table. *Journal of the American Statistical Association*, **64**, 1323–1341.

Ito, K. (1969) On the effect of heteroscedasticity and non-normality upon some multivariate test procedures, in *Multivariate Analysis II* (ed. P.R. Krishnaiah). Academic Press, New York.

John, S. (1971) Some optimal multivariate tests. *Biometrika*, **58**, 123–127.

Joreskog, K.G. (1970) A general method for analysis of covariance structures. *Biometrika*, **57**, 239–251.

Joreskog, K.G. (1973) Analysis of covariance structures, in *Multivariate Analysis III* (ed. P.R. Krishnaiah). Academic Press, New York, pp. 263–285.

Joreskog, K.G. (1978) Structural analysis of covariance and correlation matrices. *Psychometrika*, **43**, 443–477.

Kabe, D.G. (1975) Some results for the GMANOVA model. *Communications in Statistics – Theory and Methods*, **4**, 813–820.

Kalbfleisch, J.D. and Prentice, R.L. (1980) *The Statistical Analysis of Failure Time Data*. Wiley, New York.

Kariya, T. (1978) The general Manova problem. *Annals of Statistics*, **6**, 200–214.

Karnofsky, D.A. and Burchenal, J.H. (1949) In *Evaluation of Chemotherapeutic Agents* (ed. C.M. MacLeod). Columbia University Press.

Koch, G.G., Freeman, J.L. and Lehnen, R.G. (1976) A general methodology for the analysis of ranked policy preference data. *International Statistical Review*, **44**, 1–28.

Koch, G.G. (1977) The interface between statistical methodology and statistical practice. *Social Statistics Section, Proceedings of the American Statistical Association*, 205–214.

Koch, G.G. and Bhapkar, V.P. (1980) Chi-square tests, in *Encyclopaedia of Statistical Sciences* (eds N.L. Johnson and S. Kotz). Wiley, New York.

Korin, B.P. (1972) Some comments on the homoscedasticity criterion M and the multivariate analysis of variance tests T^2, W and R. *Biometrika*, **59**, 215–216.

Kres, H. (1983) *Statistical Tables for Multivariate Analysis*. Springer Verlag, New York.

Kruskal, W.H. and Wallis, W.A. (1953) Use of ranks in one criterion variance analysis. *Journal of the American Statistical Association*, **46**, 583–621.

Laird, N.M. (1978) Non-parametric likelihood estimates of a mixing distribution. *Journal of the American Statistical Association*, **73**, 805–811.

Landis, J.R., Heyman, E.R. and Koch, G.G. (1978) Average partial association in three way contingency tables: A review and discussion of alternative tests. *International Statistical Review*, **46**, 237–254.

Lawless, J.F. (1982) *Statistical Models and Methods for Lifetime Data*. Wiley, New York.

Lawley, D.N. (1938) A generalisation of Fisher's Z-test. *Biometrika*, **30**, 180–187.

Lindley, D.V. and Smith, A.F.M. (1972) Bayes estimates for the linear model (with discussion). *Journal of the Royal Statistical Society*, B, **34**, 1–41.

Lindman, H.R. (1974) *Analysis of Variance in Complex Experimental Designs*. W.H. Freeman, London.

Lindsay, B.G. (1983a) The geometry of mixture likelihoods: a general theory. *Annals of Statistics*, **11**, 86–94.

Lindsay, B.G. (1983b) The geometry of mixture likelihoods, Part II: The exponential family. *Annals of Statistics*, **11**, 783–792.

McCullagh, P. (1983) Quasi-likelihood functions. *Annals of Statistics*, **11**, 59–67.

McCullagh, P. and Nelder, J.A. (1983) *Generalized Linear Models*. Chapman and Hall, London. (2nd edition 1989.)

McDonald, R.P. (1978) A simple comprehensive model for the analysis of covariance structures. *British Journal of Mathematical and Statistical Psychology*, **31**, 59–72.

McDonald, R.P. (1980) A simple comprehensive model for the

analysis of covariance structures: some remarks on applications. *British Journal of Mathematical and Statistical Psychology*, **33**, 161–183.

MacDougall, J.D., Wenger, H.A. and Green, H.J. (1982) *Physiological Testing of the Elite Athlete*. Mutual Press Ltd.

Magnus, J.R. (1978) Maximum likelihood estimation of the GLS model with unknown parameters in the disturbance covariance matrix. *Journal of Econometrics*, **7**, 281–312.

Mainland, D. (1963) *Elementary Medical Statistics*, 2nd edn. W.B. Saunders, London.

Mantel, N. (1963) Chi-square tests with one degree of freedom: extensions of the Mantel–Haenszel procedure. *Journal of the American Statistical Association*, **58**, 690–700.

Mantel, N. and Byar, D.P. (1978) Marginal homogeneity, symmetry, and independence. *Communications in Statistics*, A, **7**, 953–976.

Mardia, K.V., Kent, J.T. and Bibby, J.M. (1979) *Multivariate Analysis*. Academic Press, New York.

Mead, R. (1970) Plant density and crop yield. *Applied Statistics*, **19**, 64–81.

Mead, R. (1989) The non-orthogonal design of experiments. *Journal of the Royal Statistical Society*, A, **152**,

Miller, R.G. (1981) *Simultaneous Statistical Inference*, 2nd edn. Springer-Verlag, New York.

Morgan, B.J.T. (1984) *Elements of Simulation*. Chapman and Hall, London.

Morrison, D.F. (1976) *Multivariate Statistical Methods*, 2nd edn. McGraw-Hill, New York.

Mukherjee, B.N. (1976) Techniques of covariance structural analysis. *Australian Journal of Statistics*, **18**, 131–150.

Nam, J. (1971) On two tests for comparing matched proportions. *Biometrics*, **27**, 945–959.

Nelder, J.A. and Wedderburn, R.W. (1972) Generalized linear models. *Journal of the Royal Statistical Society*, A, **135**, 370–384.

Nelder, J.A. (1985) Quasilikelihood and GLIM, in *Generalized Linear Models* (eds R. Gilchrist, B. Francis and J. Whittaker). Springer-Verlag, New York, pp. 102–127.

Olsen, C.L. (1974) Comparative robustness of six tests in multivariate analysis of variance. *Journal of the American Statistical Association*, **69**, 894–908.

Olsen, C.L. (1976) On choosing a test statistic in multivariate analysis of variance. *Psychological Bulletin*, **83**, 579–586.

Orchard, T. and Woodbury, M.A. (1972) A missing information principle: theory and applications, in *Proceedings of the Sixth Berkeley Symposium on Mathematical Statistics and Probability*, Vol. 1 (eds L.M. Le Cam, J. Neyman and E.L. Scott). University of California Press, pp. 697–715.

Pillai, K.C.S. (1955) Some new test criteria in multivariate analysis. *Annals of Mathematical Statistics*, **26**, 117–121.

Plackett, R.L. (1974) *The Analysis of Categorical Data*. Griffin, London.

Prentice, R.L. (1986) Binary regression using an extended beta-binomial distribution with discussion of correlation induced by covariate measurement errors. *Journal of the American Statistical Association*, **81**, 321–327.

Quade, D. (1967) Rank analysis of covariance. *Journal of the American Statistical Association*, **62**, 1187–1200.

Racine-Poon, A., Grieve, A.P., Fluhler, H., and Smith, A.F.M. (1986) Bayesian methods in practice: experiences in the pharmaceutical industry (with discussion). *Applied Statistics*, **35**, 93–150.

Rao, C.R. (1946) Tests with discriminant functions in multivariate analysis. *Sankhya*, **7**, 407–414.

Rao, C.R. (1948) Tests of significance in multivariate analysis. *Biometrika*, **35**, 58–79.

Rao, C.R. (1949) On some problems arising out of discrimination with multiple characters. *Sankhya*, **9**, 343–366.

Rao, C.R. (1950) A note on the distribution of $D_p^2 \epsilon - D_p^2$ and some computational aspects of D^2 statistics and discriminant function. *Sankhya*, **10**, 257–268.

Rao, C.R. (1959) Some problems involving linear hypotheses in multivariate analysis. *Biometrika*, **46**, 49–58.

Rao, C.R. (1965) Linear statistical inference and its applications. Wiley, New York.

Rao, C.R. (1966) Covariance adjustment and related problems in multivariate analysis, in *Multivariate Analysis II* (ed.P.R. Krishnaiah). Academic Press, New York, pp. 87–103.

Rao, C.R. (1967) Least squares theory using an estimated dispersion matrix and its application to measurement of signals. *Proceedings of the 5th Berkeley Symposium on Mathematical Statistics and Probability*, Vol. I, pp. 355–372.

Rao, C.R. (1972) Recent trends of research work in multivariate analysis. *Biometrics*, **28**, 3–22.

Reinsel, G. (1984) A note on conditional prediction in the multivariate linear model. *Journal of the Royal Statistical Society*, B, **46**, 109–117.

Ripley, B.D. (1978) *Stochastic Simulation.* Wiley, New York.

Rogers, G.S. and Young, D.L. (1977) Explicit maximum likelihood estimators for certain patterned covariance matrices. *Communications in Statistics – Theory and Methods,* A, **6**, 121–133.

Roy, J. (1958) Step-down procedure in multivariate analysis. *Annals of Mathematical Statistics,* **29**, 1177–1187.

Roy, S.N. (1953) On a heuristic method of test construction and its use in multivariate analysis. *Annals of Mathematical Statistics,* **24**, 220–238.

Roy, S.N., Gnanadesikan, R. and Srivastava, J.N. (1971) *Analysis and Design of Certain Quantitative Multiresponse Experiments.* Pergamon Press.

Scheffe, H. (1959) *The Analysis of Variance.* Wiley, New York.

Schweder, T. and Spjotvoll, E. (1982) Plots of *p*-values to evaluate many tests simultaneously. *Biometrika,* **69**, 493–502.

Searle, S.R. (1971) *Linear Models.* Wiley, New York.

Seber, G.A.F. (1984) *Multivariate Observations.* Wiley, New York.

Self, S.G. and Liang, K.Y. (1987) Asymptotic properties of maximum likelihood estimators and likelihood ratio tests under nonstandard conditions. *Journal of the American Statistical Association,* **82**, 605–610.

Sen, P.K. (1968) On a class of aligned rank order tests in two-way layouts. *Annals of Mathematical Statistics,* **39**, 1115–1124.

Share, D.L. (1984) Interpreting the output of multivariate analyses: a discussion of current approaches. *British Journal of Psychology,* **75**, 349–362.

Siddiqui, M.M. (1958) Covariance of least-squares estimates when residuals are correlated. *Annals of Mathematical Statistics,***29**, 1251–1256.

Simes, R.J. (1986) An improved Bonferroni procedure for multiple tests of significance. *Biometrika,* **73**, 751–754.

Singer, B. (1979) *Distribution-free Methods for Non-parametric Problems: A Classified and Selected Bibliography.* The British Psychological Society, Leicester.

Smith, A.F.M. (1973) A general Bayesian linear model. *Journal of the Royal Statistical Society,* B, **35**, 67–75.

Smith, P.L. (1979) Splines as a useful and convenient statistical tool. *American Statistician,* **33**, 57–62.

Somes, G.W. and Bhapkar, V.P. (1977) A simulation study on a Wald statistic and Cochran's *Q* statistic for stratified samples. *Biometrics,* **33**, 643–651.

Steel, R.G.D. and Torrie, J.H. (1980) *Principles and Procedures of Statistics: A Biometrical Approach*, 2nd edn. McGraw-Hill, New York.

Stevens, J.P. (1980) Power of the multivariate analysis of variance tests. *Psychological Bulletin*, **88**, 728–737.

Stevens, W.L. (1951) Asymptotic regression. *Biometrics*, 7, 247–267.

Sugiura, N. (1972) Locally best invariant test for sphericity and the limiting distributions. *Annals of Mathematical Statistics*, **43**, 1312–1316.

Tatsuoka, M.M. (1971) *Multivariate analysis: Techniques for Educational and Psychological Research*. Wiley, New York.

Teeter, R.A. (1982) *Effects of Measurement Error in Piecewise Regression Models*. Ph.D. dissertation, Department of Biostatistics, UNC at Chapel Hill.

Timm, N.H. (1975) *Multivariate Analysis with Applications in Education and Psychology*. Brooks/Cole Publishing Company, Monterey, California.

Tuma, N.B. and Hannan, M. (1985) *Social Dynamics*. Academic Press, New York.

Turner, M.E., Monroe, R.J. and Lucas, H.L. (1961) Generalised asymptotic regression and nonlinear path analysis. *Biometrics*, **17**, 120–143.

Wald, A. (1943) Tests of statistical hypotheses concerning several parameters when the number of observations is large. *Transactions of the American Mathematical Society*, **54**, 426–482.

Wedderburn, R.W.M. (1974) Quasi-likelihood functions, generalized linear models and the Gauss–Newton method. *Biometrika*, **61** 439–447.

Whittle, P. (1961) Gaussian estimation in stationary time series. *Bulletin of the International Statistical Institute*, **39**, 1–26.

Wilks, S.S. (1932) Certain generalisations in the analysis of variance. *Biometrika*, **24**, 471–494.

Williams, D.A. (1982) Extra-binomial variation in logistic linear models. *Applied Statistics*, **31**, 144–148.

Winer, B.J. (1971) *Statistical Principles in Experimental Design*, 2nd edn. McGraw-Hill, New York.

Worsley, K.J. (1988) Spatial correlation of tomographic images of the human brain. (Submitted.)

Author index

Subject index